HOLISTIC FIRE STRATEGIES

The search for a global methodology

Paul Bryant PhD.

DEDICATION

I dedicate this book to my daughters, Ruby and Poppy, and my family, friends, and work colleagues.

I completed this book during the Covid19 pandemic lockdown of 2020 and 2021 in Sitges, Spain. A strange time! Accordingly, I would also like to dedicate this book to frontline workers around the world. Those who risked their lives to save ours – from doctors and nurses to bus drivers and supermarket workers. Thank you.

And finally, I dedicate this book to the person who originally inspired me to develop holistic fire strategies following my meetings with her in 2016. Thank you, Adreena Parkin-Coates.

Paul Bryant

Contents

Paul Bryant

Figures

Paul Bryant

Tables

Paul Bryant

ACKNOWLEDGMENTS

Thanks to Professor Dorota Brzezinska for the assistance and support she has offered in the preparation of this book, which reflects my PhD thesis, "A semiquantitative method for the evaluation of holistic fire strategies for non-public buildings." Without her, I doubt this research would have evolved in the way it has. I would also like to thank the staff, fellow students, and professors, including visiting professors, at the *Politechnika Łodzka*, Poland, for being so supportive. This led to the defence of my thesis in June 2021 at their Faculty of Civil Engineering. I also acknowledge the efforts of Janusz Januszewski, who helped initiate my relationship with Poland.

I want to thank the following for their technical contributions to the national codes analysis (in alphabetical order): Salim Al Ali (UAE), Bob Docherty (UK), Luca Fiorentini (Italy), Charlie Fleischmann (New Zealand), Yousef Ghasemzadeh (Iran), Aaron Johnson (USA), Kelvin Wong & Stephen Xu (China).

Thanks too to my family and friends in the UK, Catalonia and Florida for supporting me and helping me with research and editing.

Finally, thanks to those who have inspired and supported me throughout my career at the Fire Offices Committee, Loss Prevention Council, London Underground, Kingfell, and Fire Cubed (Nicole and Tony). I also want to include those who have supported me via some well-loved UK fire associations. These include the IFSM, IFPO, FIA, NAHFO, and the Worshipful Company of Firefighters.

Every time we suffer a negative event, at least one outcome will be positive. It may be a lesson, an opportunity, or simply a new way of thinking.

Every time we try to hide a negative event, not only will we not learn but we will prevent others from learning.

Every time we ignore a negative event, that event becomes the new normal.

Paul Bryant

So. Are you telling me that the beautiful high rise complex I designed in Abu Dhabi can't be built here because you have different fire codes? For real?

ARCHITECT

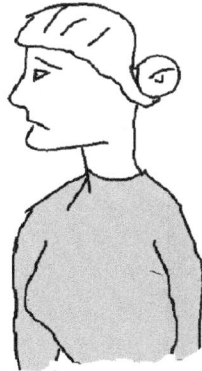

FIRE ENGINEER *PB*

I find you guilty of gross professional negligence. You failed to note the design flaw in section 28.3 of the fire strategy before approving it. TAKE HIM DOWN!

JUDGE

Gulp!

ENFORCEMENT FIRE ENGINEER *PB*

Preamble

The notion behind the development of a holistic and global approach to fire strategy preparation was conceived in 2017, shortly after discussions with the UK fire authority. The outcome was a belief that a more rigorous and consistent approach was now necessary, bearing in mind that buildings were becoming more complex and sophisticated. An initial hypothesis, developed way back in 1996, questioned whether the theory and practice behind the formulation of business strategies could be applied to the world of fire engineering. This idea followed studies leading towards an MBA from City University Business School in London, completed in 1993. At that time, business theory and strategy were influenced mainly by American expertise, with the natural home of business strategy resting at Harvard Business School, renowned as one of the top establishments for the subject in the world.

A key leader in the development of business strategies is Michael E Porter. Michael is credited as the founder of modern business strategy and is described as one of the world's most influential thinkers on management and competitiveness. He is the author of 19 books and over 130 articles. He is currently the Bishop William Lawrence University Professor at Harvard Business School and a director of the school's Institute for Strategy and Competitiveness, founded in 2001.

One of his earliest books covered competitive strategy in business[1]. In this publication, he had created a series of analytical models to evaluate and compare business strategies. These tools enabled the development of key features and requirements for a specific business profile.

Porter's ideas are a vital area of study for the vast majority of MBA (Master of Business Administration) courses across many countries. One of the main features of his work was in the development of pictorial models to assist in the analysis of businesses. These models used a range of techniques to illustrate aspects of business strategy that could not readily be understood by verbal description. I believed that similar models could be created to assist in the development of fire strategies. Early concepts were developed and published in 1996 by a UK fire safety journal[2] over two issues. The articles were titled "Fire protection strategies – Parts 1 and 2". There was a modicum of interest by the fire safety community in the ideas - but not much more.

It was a decade later when the opportunity arose to author a British Standard Specification for fire strategies. The British Standard Specification - PAS 911: Fire strategies – guidance and framework for their formulation[3] was published in 2007. The release of this PAS was around the same time as a new British Standard, BS 9999[4], designed to provide an improved prescriptive approach to fire safety. In addition, the Standard introduced the concept of risk profiles.

I introduced early fire strategy development models, together with new ideas, into PAS 911. Note that the PAS explicitly stated that it could be used for various technical standards and could apply in any country practising fire engineering.

By this time, the use of performance-based fire safety engineering methods had already matured. Standards providing a recommendatory framework for this approach were available. One of these was British Standard BS 7974[5], supported by a suite of "drafts for development" or DDs. These DDs guided professional fire engineers to derive performance-based engineered solutions, thus giving them much greater freedom to apply their ideas while still following a British Standard.

Greater flexibility did not imply, conclusively, that there were no challenges ahead. The most notable of these was that the auditing of performance-based fire strategies was made more difficult by the added complexity and variability in approach. Those tasked with approval must have both the competency to fully understand the decisions made by the project fire engineers and be assured that the engineers were also suitably competent. Slight changes in the early stages of a building design could dictate whether additional fire safety and protections provisions were necessary - or not. Even though fire engineers generally have codes to follow, these may be advisory, allowing engineers to cherry-pick recommendations as they so wished.

It is great to see a city enjoying a construction boom. New, exciting, and unique building profiles will proliferate. Good news - Possibly? Ensuring buildings are *firesafe* – a problem? A problem shared by enforcement authorities and agencies throughout the international built environment. Every fire strategy submitted may differ in style, content, methods adopted, and assumptions made. Some make use of modelling techniques to determine appropriate evacuation times, smoke control parameters, etc. Many do not.

Each fire strategy typically requires intense scrutiny by appropriately qualified and experienced persons. So, naturally, there could, and should, be a certain degree of reliance on the abilities of fire engineers involved in the preparation of the strategies. Nevertheless, the enforcement authority must have sufficient assurance in signing off a strategy that is fit for purpose. And, of course, fire safety is easy pickings for national budgetary cuts.

The question arose in 2017 as to whether there could be another way forward – a solution that will regulate the fire strategy formulation and, more importantly, evaluation process? Could such a solution be applied on a global basis?

One problem here was, and is, that each country had its own set of fire safety rules, regulations and codes, and its methods of managing building programmes. A simple global solution is not apparent. One chapter of this book covers this in more detail. There is a need to improve the auditability by perhaps providing a more consistent framework wherever the building is located. The concept should ensure that enforcement authorities could review a fire strategy by quickly identifying the thought process and structure, the assumptions made, and the methodology used.

To ensure that the system is available to all, it would make sense to utilise the internet. Perhaps it should be accessible via a web-based portal, accessible by the project teams and enforcement authorities? The complete process could be online – avoiding the need for paper but allowing for paper versions if so wanted.

Another feature of many fire strategies is the narrow focus of formulation. Many fire engineers simply set out to ensure that the fire safety design solution is compliant with the relevant codes. The problem with this approach is that it does not encourage a proper degree of *holistic* thinking. It is easy for a fire engineer to be blinkered by the regulations and codes they use to achieve compliance. There is a need to encourage them to think about *what* they are doing *when* they are doing it. If they religiously apply the stated requirements, they are likely to be focusing on a fire safety design solution focused on the perspective of life safety. Most fire engineers will not purposefully consider property and asset protection issues as additional objectives unless they are specifically told to do so. Furthermore, business continuity and environmental protection objectives are becoming increasingly more critical yet are often ignored.

Some strategy documents explicitly state that *"this fire strategy does not cover extreme events"*. What does this mean? Is it because fire engineers cannot guarantee that their designs will save the building and the people in that building? Note that fire safety legislation may assume that any event other than a single fire event could be considered an extreme event in any section of a building. The opportunity to undertake a proper and thorough threat analysis could help identify possible fire scenarios that may otherwise have been ignored.

I hope you enjoy the book, even if you do not agree with some of my ideas.

1 Is there a need for change?

"If it ain't broke, don't fix it..." – a declaration that unintentionally discourages innovation.

Some may say that the subjects of fire safety, fire protection, and fire engineering are sufficiently mature that there is no need to seek fundamental improvement. Fires affecting most building profiles are not commonplace, although residential fires continue to outstrip other building fires in terms of frequency. Fatalities caused by fire or by the effects of fire are relatively rare. But when there are multiple fatalities, the news quickly spreads around the world.

UK fire statistics have shown a general decline in fire incidents over the last two decades[6], including both accidental fires and fires caused by deliberate causes such as arson. A similar trend is found in the United States[7].

Nonetheless, the detail behind the trends does not naturally infer success. Although deaths caused by fire appear to be decreasing (roughly in line with the decrease in fire incidents), if we take a broader look at the overall cost of fire, there are some additional lessons to be learned. For instance, according to the US National Fire Protection Association (NFPA), fires in the USA in 2017 caused approximately $10.7 billion[8] in property damage. There is no sign of this figure declining over time. Even after the value is adjusted for inflation and other costs, the problem has not been adequately tackled. There is more to do.

When it comes to business, the impact of a fire can often be disastrous. US Insurer NFU[9] highlight that 80% of companies fail within 18 months of a major fire incident. Safety Management Journal[10] put this figure at 70% (even though they state that an exact figure is hard to determine). In either case, the statistic is high and is unlikely to reduce unless businesses heed the warnings and take appropriate actions.

Then there is the environmental damage caused by a fire. When thinking about this, we may revert to the news coverage of the instances of wildfires increasingly reported, particularly in the US and Australia. These incidents' direct and indirect costs will be in the billions of dollars, but this does not consider the cost to the planet itself. Environmental damage due to a building fire is often far less publicised and possibly still not fully understood. Examples of this are given later in this book.

Perhaps the status quo *viz a viz* fire safety may require re-examination, not just in our attitude towards fire safety but also in how fire safety is specified, managed, and audited.

In my book "Fire strategies – strategic thinking"[11], I describe the subject of fire engineering as a "black art". An unfair description of a subject that has amassed a wealth of knowledge? Is there not an abundance of codes, standards, rules, specifications throughout most nations that determine how we make our buildings *firesafe*? Is this not backed up by extensive research carried out in all corners of the globe? Is not fire safety a cornerstone of most building legislation?

With so much in the way of national and international guidance, it could be argued that opinion plays little in developing a fire strategy. After all, the fire engineer follows a set of rules covering means of escape, the use of active and passive fire protection, facilities for firefighters, and so on. Simple – but possibly restrictive?

Fire engineers recognised the limitations of following prescriptive fire safety rules decades ago. Nevertheless, such fire safety rules have been with us from the early days of standardisation. Significant revisions of these rules are either due to new lessons learned by the scientific and engineering communities or a major fire.

Rules for fire safety of buildings date back centuries. The protection of people against the worst ravages of fire is a principle that has been around probably as long as civilised society itself.

Typically, the rules support the relevant national fire safety legislation. Prescriptive rules would provide *absolute* guidance such as travel distance to a place of safety, the location, extent, and rating of active and passive fire protection, and means for firefighting. Consequently, new and innovative building design would be severely restricted.

Performance-based approaches, which allow for greater flexibility in the application of fire safety, have become the *go-to* solution over the last few decades for more complex building arrangements. Without such progress, it could be argued that the more ambitious building designs we see today would never have got off the ground. The approach allows a greater degree of diversity in design, assumptions made at the outset, and methods used to arrive at a fire-engineered solution. I introduced the term "opinion engineering" in my book "fire strategies – strategic thinking" with mixed reactions. If we seek a method that could improve the *consistency* of approach, would this not limit the *subjectivity* of approach?

There is another issue that may not be instantly apparent to those outside of fire safety enforcement.

Enforcement authorities[12] have highlighted challenges concerning the receipt of fire strategy documents. Typically, every submitted fire strategy may differ in the degree of detail, as well as format. Some strategies adopt a pure performance-based approach using CFD modelling. Some tend to bridge the gap between prescription and performance setting. Some are ambiguous as to the assumptions and decisions made. All of this introduces a dilemma for those tasked with ensuring that the fire strategies are, indeed, good enough. It was a much simpler task when strategies used prescription. Now the task had become much more difficult.

Moreover, building design itself had become much more complex, introducing further frustration for enforcement authorities who typically cannot afford the resource and cost required to analyse the strategies properly. The same issues can be found worldwide, with enforcement authorities facing similar dilemmas, especially where building projects are increasing at a high rate.

Surely there could be a methodology that would allow fire engineers to follow a process that would vastly improve the consistency of their output. Given that the issue is a global problem, then could not the process also be global? After all, even acknowledging national conditions, building designs are increasingly much the same around the world? Therefore, the approach should be holistic.

This Book introduces principles that are not readily included in fire engineering designs:

- Enhanced Objectives Setting.

- Threat Analysis.

One of the conclusions from the fire statistical trends raised earlier is that objectives outside of life safety remain a concern. National legislation tends to focus on life safety so that fire strategies would be bound primarily towards the requirements for life safety. However, some strategies would venture outside this constraint - to consider other goals such as property protection, possibly following an instruction from stakeholders such as insurers. Preparing a fire strategy for a new building is often a golden opportunity to undertake a complete and thorough review of all potential objectives at the onset of a construction project.

I first raised the idea of four primary objectives and sixteen sub-objectives in British Standard Specification PAS 911[13]. The primary objectives are life safety, property protection, business protection & continuity, and environmental protection. To a greater or lesser extent, all four primary objectives may be relevant for each fire strategy, in whatever sector, and in any region or country.

The concept of introducing threat analysis is in recognition that our social environment is changing. New threats could lead to new fire scenarios that may not have been adequately considered in the past. A fire-engineered solution should incorporate a proper consideration of threats leading to a fire that could impact the building, its occupancy, and its processes.

Most fire strategies adopt the concept of a single fire event within the building boundary at any one time. When applying their national codes to a new building project, it is doubted whether some fire engineers are even aware of that fact. Some codes accept the principle that anything other than this event may be considered an extreme event. It is not uncommon for fire strategies to state that "extreme events" are not covered. We may believe that an extreme event is something like a massive explosion. But as commented on above, two independent simultaneous fires in different parts of a building may be considered extreme. Is this acceptable to all the stakeholders? Probably not, so there is ample good reason to consider threat analysis within a thorough examination of all fire safety features.

When these additional considerations are combined, we have a globally based, all-encompassing approach to fire engineering. The term *Holistic Fire Strategies* seems to be an appropriate title for this idea.

Given the current wealth of knowledge about the subject of fire, the concepts described in this book are not intended to change fire science or the application of fire engineering principles. Instead, the fundamental principles and objectives are to provide a highly auditable framework. The primary beneficiaries of this will be the enforcement authorities, as this will increase their confidence in holistic fire strategies and save them time and money. Nonetheless, we need to consider all stakeholders and if such a system could benefit each of them. This idea is covered later in this book.

Let us work on the following key principles for a suggested holistic approach:

a)   To ensure that a fire-engineered solution properly accounts for the real and perceived threats affecting the building, occupancy, and processes. Extreme events may or may not be included based upon a risk and scenario evaluation.

b)   That we consider, fully, all objectives, not just those applicable to national regulations. Note that comparison with national rules should be included within the process.

c)   We utilise all existing recognised means to develop holistic fire strategies.

d)   Critical to holistic fire strategies is that a measurement system controls the analysis and design process to allow full auditability and comparison at any stage of the process. Consequently, this will provide third parties with greater assurance that the solution is compliant with "holistic fire strategy" metrics.

e) The process and metrics must be transferable globally, such that they will be the same wherever they are applied. The concept will have to incorporate assessment and comparison with national requirements.

f) The framework and process cover feasibility through to delivery of the holistic fire strategy.

This book will gently take the reader through my ideas for the *holistic fire strategy* methodology. It will introduce concepts that I believe are necessary for a mature but still developing industry. It is doubtful that the reader will agree with everything I propose. The ideas within should be treated as a work in progress.

Paul Bryant

2 Lessons from the past

Lessons from the past are indicators for the future.

If there is a *want* to change the direction of fire strategy formulation, or indeed if it is thought that there is *no* need, then we must understand the steps that have led us to now. I will use the United Kingdom to provide one example of the progression of standards development for fire safety. No doubt, many countries will have a similar tale to tell. In chapter 4, I will explore the international situation.

The United Kingdom, as with most parts of Europe, had used stone masonry for centuries to create their primary buildings, such as castles, churches, country houses, and so on. These building methods were acknowledged as providing inherent fire protection by separating one section of a building from another, such that a fire could be relatively easily contained. This strategy would give the best chances of fighting the fire before it could engulf the whole building. Nonetheless, the same building style was not extended to most smaller buildings throughout Britain at the time. This was evident following the Great Fire of London in 1666. The fire destroyed 13,200 houses, 87 parish churches, St Paul's Cathedral, and most of the buildings of the City of London.

An article prepared for IFSEC Global[14] provides key lessons from that fire. These included:

Lack of passive fire protection: At the time, London consisted of a jumble of medieval streets with wooden, thatched houses. The risks associated with thatched roofing were already well known, and consequently, such construction methods were officially prohibited. But these laws were, overall, ignored. Furthermore, six and seven-storey timbered tenement dwellings were expanded on the upper floors to form overhanging jetties in narrow streets. The construction enabled a fire to jump to neighbouring houses quickly. Note that these jetties were forbidden by proclamation by the King (Charles II) at the time.

Combustibles: Open fireplaces, candles, etc., were commonplace in such houses. Many occupiers stored gunpowder. Some of the buildings were also used to include foundries, smithies, glaziers, etc.

Firefighting: *Trained Bands* were persons who patrolled the streets at night for fires and other potential emergencies. Equipment such as axes and firehooks were made available for pulling down buildings. Laws required Parish churches to keep long ladders for accessing towers. Early versions of our relatable fire engine (or fire truck) were also available. These proved to be unusable in many cases. Problems included the long distances to get to a fire, limited reach, no hoses, and the inability to access narrow streets due to their width.

The fire revealed inadequacies in the rules for fire safety. As a result, the King introduced legislation that buildings should no longer be built of wood and that roadways should be widened to reduce the risk of fires jumping from one block to the next. Consequently, fire safety requirements and regulations were introduced.

Statutory provisions continued to evolve, often because of fires that killed considerable numbers of people. In the 19th century, detailed stipulations were made for the safety of people within premises in a fire situation. These were swiftly taken up by many parts of the world, particularly in countries that were regarded as part of the "first world" and those who were part of the British Empire. Before this, large-scale, ad-hoc tests were carried out in the mid-18th century with the recognition that fire should be confined to the room of origin rather than the building. It was not until the end of the 19th century that standard fire tests were used.

Together with the accompanying rules, British fire safety legislation tended to develop piecemeal, following significant fires affecting a range of building profiles. Legislation relevant to fire safety included The Explosives Act (1875); The Petroleum (Consolidation) Act (1928); The Factories Act (1937); The Fire Services Act (1947); The Factories Act (1961); The Licensing Act (1961) and The Offices, Shops, and Railway Premises Act (1963).

In 1971, the UK's Fire Precautions Act replaced previous relevant legislation. One of the primary features of this Act was that building occupiers were required to hold a fire certificate. This certificate described the range of fire safety provisions incorporated within the building, often accompanied by annotated plans. Authorities had the option of closing premises where it was believed that occupants were at risk. In a way, these certificates were not dissimilar to the objectives inherent in a fire strategy document we now use.

But major fires continued. On each occasion, legislation was introduced or reviewed. For example, in Manchester, UK[15], a fire in a retail department store in May 1979 led to a death toll of ten, with many others seriously injured. The subsequent investigation found that most of those who died were in the restaurant above the store. The subsequent investigation revealed that the smoke was so thick that shoppers could not find their way to the exits. Note that iron bars were installed on the windows, preventing escape or rescue via these access points. Although there was a specific Act in force at the time of the fire (The Fire Precautions (Factories, Offices, Shops, and Railway Premises) Regulations 1976), the store in question did not have a fire certificate. The owners were in the process of making improvements. The fire also prompted calls for sprinkler systems in department stores as well as better staff training.

Two other major fires led to new legislation. One was the Bradford Football Stadium fire in May 1985, when 56 spectators died and upwards of 265 injured. This incident paved the way for the Fire Safety and Safety of Places of Sport Act (1987).

There was one major fire that probably had more impact on my career than any other event. That was the King's Cross Underground Station fire on 18 November 1987, which resulted in the loss of 31 people in horrific circumstances. The Fire Precautions (Sub-surface Railway Stations) Regulations were published in1989. Soon after I arrived at London Underground in 1991, I was put in charge of their fire engineering division. We were involved in putting together a range of fire safety measures and active and passive fire protection systems for around 115 sub-surface railway stations (approximately half of all London Underground Stations). During this time, I understood the nuances and benefits of integrating management and system concepts to form a unified and complementary fire strategy.

Moving forward in time, to 2005, the UK introduced legislation that avoided the continual need to catch up with the impact of new fire-related disasters. The UK Government's Regulatory Reform (Fire Safety) Order 2005 was enforced from 2006. For the first time in the UK fire safety history, a proactive approach was adopted to ensure appropriate levels of fire safety are applied. The legislation required the use of fire safety risk assessments. Fire safety provisions and precautions could then be applied based upon the perceived fire risk. This had one drawback: it relied upon the decisions made by a "fire risk assessor". Assessors often had different opinions, typically based upon their experiences and qualifications, despite the availability of a set of national guidance documents. Third-party certification schemes were set up to increase professionalism in the undertaking of fire risk assessments. Nevertheless, opinion is still a key ingredient in a fire risk assessment. Is there another way of improving the consistency of the approach?

The fire risk assessment process primarily applies to existing operational buildings. For new or modified buildings, the legislative requirements are typically supported by the UK Building Regulations and specifically Approved Document B (ADB), which covers fire safety[16]. Note that this document is provided in two volumes, one for dwellings and the other for non-dwellings. ADB further divides requirements up into five parts:

B1: Means of warning and escape

B2: Internal fire spread (linings)

B3: Internal fire spread (structure)

B4: External fire spread

B5: Access and facilities for the fire service

British Standards Institution had published a range of fire safety standards to support legislation for many decades. Traditionally, it was the BS 5588 series of standards that specified overall fire safety requirements for buildings. The series title was *Fire precautions in the design, construction, and use of buildings*. I say "was" as this whole series of standards has mostly been withdrawn in favour of BS 9999, which I will introduce shortly. There are several other British Standards that cover most aspects of passive and active fire protection and fire safety management. Some of which are also European and international standards.

Currently. the most relevant UK standard for the specification of fire safety in buildings is BS9999[17]. The standard provides recommendations for the design, management, and use of buildings to achieve reasonable standards of fire safety for all people in and around them. It also guides the ongoing management of fire safety within a building throughout its entire life cycle, including information for designers to ensure that the overall design of a building assists and enhances the management of fire safety. More recently, a British Standard (BS9997)[18] was published to improve fire safety management techniques.

BS9999 recognised that additional flexibility is required and included the technique of risk profiling. The risk profile uses two parameters: the potential rate of fire growth and the occupancy profile (e.g., occupants' knowledge of the building, the possibility that occupants are not awake during a fire, etc.).

When it comes to consideration of objectives, BS9999 acknowledges that it primarily safeguards the lives of building occupants and firefighters. It points out that issues such as property protection, the environment, communities, and business/service viability are outside of its scope.

A UK performance-based fire safety approach was formalised with a "Draft for Development" introduced by British Standards Institution (BSI) in 1997. The standard was titled DD240-1[19]. The scope of the standard highlighted that it provided a framework for an engineering approach to the achievement of fire safety in buildings by giving guidance on the application of scientific and engineering principles to the protection of people and property from fire. A second supporting standard, DD240-2, was also published to provide commentary on the equations given in the first part.

This Draft for Development was seen as a breakthrough by allowing building designers an alternative to prescription based upon performance objectives determined for that building. The idea was that one or more meetings would be held involving relevant stakeholders, who would set appropriate performance parameters. These meetings were described as "qualitative design reviews" in that the qualitative decisions made would guide the subsequent quantitative analysis. The key concepts of DD240 were developed further by BSI when, in 2001, British Standard BS 7974[20] was published to supersede DD240. BS 7974 provides the same framework for an engineering approach as described for DD240-1. It also provides a *"rational methodology for the design of buildings"*. The standard applies to the design of new buildings and the appraisal of existing buildings. The key benefits highlighted by the British Standard were that it provided;

- the designer with a disciplined approach to fire safety design.

- safety levels for alternative designs to be compared.

- a basis for the selection of appropriate fire protection systems.

- opportunities for innovative design.

- information on the management of fire safety for a building.

The standard is supported by guidance documents, published as "Public Documents" or PDs.

These documents provide fire safety engineers with additional information, allowing them to formulate effective and relevant performance-based fire strategies. Each PD, also referred to as a sub-system, covers a specific area of consideration:

- Initiation and development of fire within the enclosure of origin (Sub-system 1);

- Spread of smoke and toxic gases within and beyond the enclosure of origin (Sub-system 2);

- Structural response to fire and fire spread beyond the enclosure of origin (Sub-system 3);

- Detection of fire and activation of fire protection systems. (Sub-system 4);

- Fire and rescue service intervention (Sub-system 5);

- Human factors. Life safety strategies. Occupant evacuation, behaviour, and condition (Sub-system 6);

- Probabilistic risk assessment (Sub-system 7);

- Property protection, business and mission continuity, and resilience (Sub-system 8).

Other than sub-system 8, the principles given ensure that the building provides suitable life safety measures to safely evacuate from a building in a fire. This is based upon two factors:

Available Safe Escape Time (ASET) represents the time duration for which safe conditions will prevail before escape routes become untenable. This period can be determined by using the sub-systems given above. It will include factors such as structural stability, protection of escape routes, and the use of active systems to prevent fire and smoke affecting the means of escape for an enhanced period, thus increasing the ASET value.

Required Safe Escape Time (RSET) represents the time necessary for occupants to safely escape from a fire, either to a final place of safety (such as outside of the building) or to a relative place of safety (such as a refuge). Clearly, in the latter case, this will require enhanced measures to ensure refuges remain safe for the required period until they can evacuate to a final place of safety. This period can be reduced by measures such as suitable fire detection and alarm systems, the type of response to the alarm, and the efficacy and sizing of horizontal and vertical escape routes. The RSET is the summation of periods from detection to alarm, pre-movement, and evacuation (Figure 1).

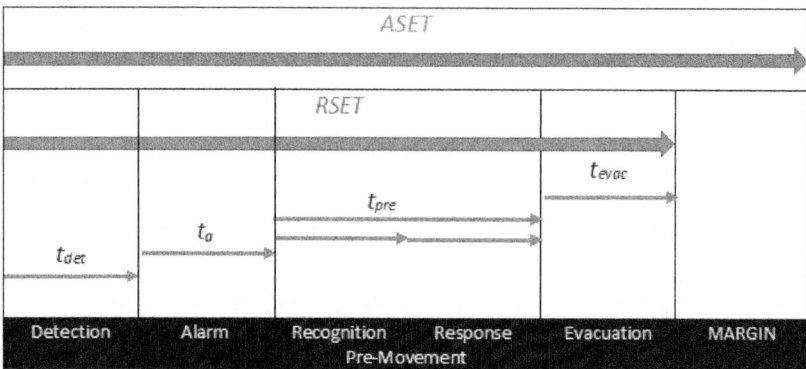

Figure 1: Breakdown of ASET & RSET Factors

The objective is that the ASET is calculated to be greater than the RSET, with a reasonable safety margin. Figure 1 provides a breakdown of the principal periods making up the RSET. Rather than calculate the above from first principles, computer models are often used to determine both conditions. Early versions made use of "zone models". These models are usually based on simple room dimensions and divide rooms or enclosures into one or more zones. A "two-zone" model, for instance, would consider an upper layer consisting of heated products of combustion and a lower layer of cooler air relatively free of combustion products. The models would allow for an assessment of the conditions over time by considering variations in the location of the fire and the expected fire size. The impact of openings in the room, such as doors or windows, can also be assessed. The models could help predict the time to detection and operation of systems such as sprinkler systems and predict the time to flashover. There are limitations in such models in that, for instance, the expected temperature within the hot layer would be the same throughout.

Field models that use sophisticated computational fluid dynamic (CFD) algorithms are more commonly used today. They work on the principles behind *Navier-Stokes* equations[21] appropriate for low-speed thermally driven flow, emphasising smoke and heat transport from fires. At any point in time, it is possible to find the temperature, velocity, and gas concentrations at any node. Such systems can help assess complex building designs and particularly 3D models. There are many modelling programmes available today. Some packages enable both ASET and RSET conditions to be modelled simultaneously in real-time. Adjustments to the criteria can even show how many persons could be affected by smoke and fumes while evacuating the building.

In this chapter, I have highlighted how legislation and standards have evolved for the UK. Similar fire safety histories will be found elsewhere. Fire safety requirements typically have developed following a tragedy. Politics and economics have doubtless played as much a role in developing fire safety legislation as the evolving scientific and engineering understanding of fire.

There have been recent and radical advancements in the regulation of fire safety. Techniques such as performance-based approaches and fire risk assessments to support fire safety legislation may seem to solve some longstanding issues. But maybe they have also introduced new challenges?

Paul Bryant

3 What is a fire strategy?

An artist needs to visualize the final picture before applying the paint.

The term fire strategy has gradually crept into the fire safety dictionary and is generally accepted and understood by the global fire community. Yet, at the time of writing this book, even Wikipedia precluded the term.

In essence, a fire strategy is simply a high-level plan, combining fire safety management, means of escape criteria, the application of active and passive fire protection, and the means to fight a fire. The idea is that the systems, layout, and management complement the overall fire safety objectives for the building, its occupants, and its processes. In some cases, elements of the strategy can be adjusted to allow for special features to be acceptable. Once the strategy has been developed, detailed designs can be prepared for fire detection and suppression systems, smoke control systems, and fire compartment layouts.

In my book "Fire strategies – strategic thinking", I compared the formulation of a fire strategy with developing business strategies and military strategies. It was the business strategy analogy that initially sparked my imagination and prompted me to develop models to help analyse fire strategies, in the same way as business strategists use evaluation "tools" to create competitive advantage.

But it was the comparison with the main facets of a military strategy that I found the most striking. The great military leaders used gained knowledge of the terrain, the groupings and movement of people, and the optimum use of resources to achieve victory. Similarly, a fire strategist must understand the nuances of the building design and arrangement (i.e., the physical terrain), the occupancy profile(s) and then apply their fire protection resources to make the building *firesafe*.

I pointed out in my book that a fire strategy is more fundamental than may be commonly understood. Let's take the example of a residential apartment. The "implicit" fire strategy will be that the building was constructed in accordance with national building regulations (all of which contain fire safety requirements). The electrical system within the apartment will include safety systems to prevent overheating or sparks igniting the furnishings. In some cases, a smoke alarm may be fitted to alert sleeping occupants of a fire. All these provisions put together amount to a fire strategy. Consequently, everything constructed for people to live, play, or work in will have a fire strategy.

I postulated that a fire strategy needs to incorporate the following five properties:

1. To be specific to the unique set of fire-related parameters of the building or structure to which it applies, the processes within, and the occupancy profiles. There is no such thing as a "generic" fire strategy.

2. To be a clear and concise document, despite the necessary and sometimes complex processes throughout its drafting. It will need to be understood by all parties affected by it and not just by fire safety professionals involved in its preparation.

3. To have the necessary level of detail to enable, for instance, fire safety management plans to be

drawn up and to provide the design criteria for passive and active fire protection. But, at the same time, it should not be so detailed that it is inflexible to changing fire safety and protection technologies or philosophies.

4. To have realistic and achievable goals relevant to the local/national fire safety requirements. A strategy will need to consider practical, logistical, and commercial limitations. A complex "all bells and whistles" strategy may be desirable from the viewpoint of a specifier or enforcer. Still, if the fire engineer cannot make it a reality, it won't be, and thus the process would mostly be a waste of time and money.

5. A fire strategy is an organic document. It should be modified and adjusted to remain true to its inherent goal, preventing and mitigating fire incidents and their impact. Drivers that dictate the need for modification of the document include changes in legislation or stakeholder requirements, revised building structures or layouts, changes in occupancy or use of the building, and newly available technology or research.

British Standard Specification PAS 911

This document was published in 2007 to provide a methodology for preparing fire strategies, whether they use prescriptive standards or a performance-based approach. This document does not give recommendations or requirements for the fire safety design of buildings. Instead, it aims to provide a consistent platform for fire strategies, such that they will follow a consistent format, whatever the building type is and wherever it is.

Historically, a fire strategy would follow national prescriptive codes or regulations, allowing for a straightforward fire strategy document. The document style made life very easy for both practitioners and those auditing their results. Prescription tends to focus on absolutes, i.e., the fire strategy either conforms to the standard or not. The advent of performance-based codes has added another way forward, although many countries continue to support prescription over pure performance-based engineering. Figure 2 shows two pie charts, one representing a prescriptive approach and the other, a performance-based approach.

Prescriptive rules are "black and white," that is, there is a right way and a wrong way. In an ideal world of prescription, there would just be black and white. It is commonly accepted that no building, or use of a building, is likely to allow for every element of a prescriptive code to be met in full unless the building is

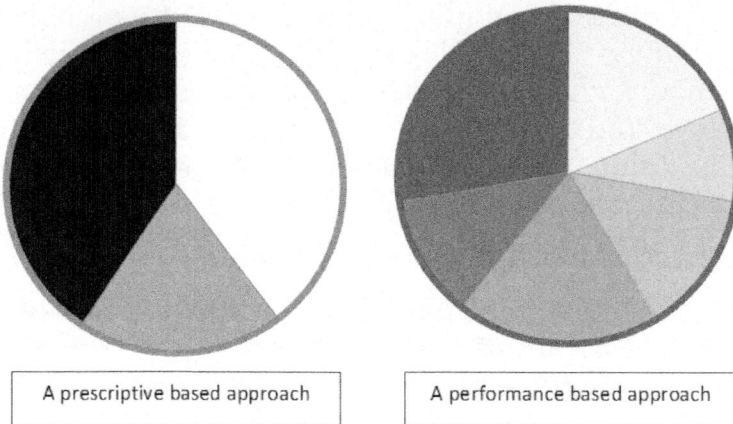

| A prescriptive based approach | A performance based approach |

Figure 2: Pie chart depiction of prescriptive and performance-based fire strategies (BS PAS 911)

highly standard in its construction and layout, as well as its use. Alternatively, the standards themselves may not be so precise as to cover every detail. There is also the possibility that the standard, or the building, has not been adequately understood, and the outcome is full compliance by default.

If a prescriptive code has been fully and thoroughly used, and the building has been appropriately assessed, there will be aspects that cannot be made to strictly comply and are thus depicted by an area of grey. These grey areas are known as variations or deviations from the standard. These areas of grey may be accepted or rejected by those who will approve the fire strategy.

A performance-based approach will not lead to any obvious right or wrong answer. Instead, a solution may be effective to a lesser or greater extent. That is, there will be degrees of grey with no absolute right or wrong approach.

I have noted over the years that many fire strategies, especially where more complex buildings are involved, are hybrid (prescriptive & performance-based). Certain aspects of the fire strategy may use a performance-based analysis, while others call upon highly prescriptive standards.

Examples could include the specification for elongated travel distances, allowed by using smoke extract systems to maintain corridor tenability to meet a performance objective. At the same time, a prescriptive code is used, for example, to determine the siting and spacing of fire detectors.

BS9999, a British Standard I introduced earlier, could be described as a hybrid standard. The use of risk profiles allows for some degree of flexibility. It recognises a trade-off between elements of structure and arrangement and the use of fire protection systems. It then calls up prescriptive standards for fire detection and alarm systems and fire suppression systems.

Given the added flexibility of BS9999, many of my ideas for a holistic fire strategy approach could possibly be met by a further enhanced version of the code.

4 An international approach?

Fire knows no boundaries.

Nations tend to support and rely on their home fire safety standards and codes. I will subsequently use the term *code* as a collective representation for rules, standards, and so on unless a specific document is referenced. Codes typically support the needs of their building regulations and fire safety legislation. Some countries may formulate codes that are unique to them. Some may take codes from other countries and modify them for their use. Some may specify and utilise foreign codes directly.

Even a national approach can be undermined when regions within one realm take differing viewpoints on fire safety standardisation. Regional variations in specific aspects of fire safety and firefighting are not unusual. Rodrigues *et al.*[22] point to the regionality of fire safety standards in Brazil. They highlight their domestic situation whereby individual states within the country have created a diversity of regulations, with varying mandatory prescriptive requirements for fire protection systems. They believe that the only way forward is for Brazil to start considering fire safety at a federal level by bringing together regional stakeholders. They also support the adoption of fire engineering techniques based upon scientific, rather than historical, determination.

Ideally, as fire engineering and scientific principles become increasingly global, fire safety specification should follow suit. International standardisation would seem to make sense. The International Standards Organisation (ISO) was formed in 1946 when delegates from 25 countries met in London and decided to create a new group *"to facilitate the international coordination and unification of industrial standards"*. They have now published over 22,000 International Standards covering many aspects of technology and manufacturing, with 164 countries taking part.

Back in the 1980s, there was an intention to formulate a range of international fire safety codes. I was involved in some of the meetings. The ISO group was known as Technical Committee 92[23]. The group had achieved some successes with a range of *International Standards* covering fire testing of building components. More recently, a series covering many aspects of fire engineering, from the use of fire zone modelling (ISO/TS 13447) to the selection of fire scenarios (ISO 16733), fire risk assessment (ISO 16732), and a core subject of this book, objectives setting (ISO/TR 16576). Sprinkler systems are also covered by an International Standard[24], although ISO standards for the many other forms of fire protection system remain elusive. Even after all the effort made by a large international group of fire experts, the national take-up of this work is not clear.

Another initiative purports that the problem with multiple differing fire codes is that there is no single authoritative way to work. The International Fire Safety Standards (IFSS) group[25] was formed to bring greater consistency by setting minimum fire safety and professionalism levels across the world. In 2018, the IFSS Coalition was launched at the UN in Geneva, Switzerland.

In the context of the IFSS Coalition's work, an international standard is established and agreed on at a global level and implemented locally.

The coalition has stated that it will provide *"universal rules that classify and define fire safety standards at project, state, national, regional, and international levels"*. They refer to research, which has shown that inconsistent approaches to the assessment and regulation of fire safety can lead to a loss of confidence by governments, financiers, investors, occupiers, and the public in buildings. They believe that, in extreme cases, this has resulted in the loss of life. Their end objective is that all higher-risk buildings to which occupiers and the public have access will publicly display a certificate of compliance with the IFSS.

It could be deduced that methodologies for performance-based fire engineering are more likely to be used on an international stage. There are two points to consider. The first is that countries are not obliged to use such guidance, even if they have taken part in their formulation. And secondly, as identified above, most countries still need to require buildings to meet their own fire safety codes, whether prescriptive or performance-based. This is unlikely to change soon. Enforcement authorities are unlikely to accept a genuinely international approach unless their national fire safety committees and other national stakeholders fully embrace this new status quo.

An excellent example of a successful global approach is the computer industry. There was a time when the industry displayed a similar degree of intransigence. There were many variants of IT-based equipment with associated operating systems, some nationally based or even corporate based. These were gradually replaced with what is now an international offering, with both hardware and software interchangeable anywhere in the world.

Are the fundamentals of fire safety standardisation so different? To help me answer this question, I used a group of international peer fire safety professionals. I posed a series of questions covering some mainstream elements of fire safety specification and analysed their responses. The countries chosen were (in alphabetical order): China, India, Iran, New Zealand, Poland, UK, and the USA. Those who assisted me are credited on the *Acknowledgements* page. I also used my experiences after a career in standards formulation and searched the internet to fill in any gaps. I apologise in advance if I have misinterpreted or omitted any of the national requirements.

Question 1: What are the most applicable fire safety codes utilised in your country?

China retains its governing building code titled GB50016-2014 (2018 edition) - Code for fire protection design of buildings (National Standard of the People's Republic of China)[26]. The Code is called up throughout China and allows for both prescriptive and performance-based approaches through expert panel review in some areas. However, evaluation methods such as risk profiling are typically not utilised. China has a separate standard for high-rise buildings titled GB 50045: Code for fire protection design of tall buildings, which covers residential buildings above 27m, and all other building uses above 24m.

Hong Kong applied its own codes before handover back to China in 1997. These were inevitably based upon requirements from the UK with local modifications. Hong Kong now has a separate local code, "Code of practice for fire safety in buildings: 2011"[27], produced by their Buildings Department for fire safety design.

India has four primary fire safety codes:

- IS 1641: Code of practice for fire safety of buildings (general): General principles and fire grading and classification.

- IS 1642 'Fire Safety of Buildings (General): Details of Construction – Code of Practice IS 1643 'Fire Safety of Buildings (General): Exposure Hazard -Code of Practice

- IS 1644: Code of practice for fire safety of buildings (general): Exit requirements and personal hazard

- IS 1646: Code of practice for fire safety of buildings (general): Electrical installations.

IS 1641 provides a classification of 9 risk profiles (incl. residential, educational, business, industrial, etc.). These are then sub-classified into types. Note that these codes were produced in the late 1980s and early 1990s. The principles of performance-based engineering were not well known at that time.

The Iran National Building Regulations: Section 3 - Fire Protection in Buildings is their most relevant Code. The document has been adapted from a combination of NFPA Codes, British Standards, and Iranian Building Codes. It has no English version and is written in their national language (Farsi). The regulations are prescriptive but allow for a performance-based approach. It is understood that a performance-based design has yet to be approved at the time of this publication. Risk profiling is generally not used for civilian applications but is for industrial applications.

New Zealand's primary fire safety requirements are covered by their Building Regulations' C Clauses. These documents are produced by their Ministry of Business, Innovation, and Employment. The sections are C1: Objectives of Clauses C2 to C6; C2: Prevention of fire occurring; C3: Fire affecting areas beyond the source; C4: Movement to a place of safety; C5: Access and safety for firefighting operations: establish the presence of hazardous substances or process in the building; C6: Structural stability. Note also that New Zealand uses a range of verification methods (VMs) and Acceptable Solutions (ASs) to support or allow modification from the C Clauses. A ten-part fire scenario assessment is also part of the verification process. I will cover this in more detail later in this book.

In Poland, their Building Regulations - Standard 256: *Polish regulations for buildings and their location* are adopted for fire safety requirements. The Regulations tend to either make use of British Standards (some of which are regulated as European standards) and relevant NFPA guidance together with BS and NFPA codes as performance-based tools. Risk profiling is currently not commonly used.

The UK Building Regulations - Approved Document B is the primary source applied as a basis of fire strategy formulation. British Standard BS 9999 is used where a degree of flexibility is necessary. As advised in the previous chapter, BS 7974 is the primary standard for a performance-based approach.

The United States primarily uses its NFPA Life Safety Code 101[28]. NFPA 101A[29], Guide on Alternative Approaches to Life Safety, allows for an engineering solution to correctly determine an equivalency methodology against NFPA 101, so allows for greater flexibility. The equivalency method uses numerically based fire safety evaluation systems with minimum mandatory values. The Codes cover a wide range of occupancy types and provide specific guidance on a chapter-by-chapter basis for different building use profiles, such as Health care, Residential board and care, Detention and correctional; Business, and Educational.

The overall impression gained from the responses is that we all like to use our own set of guidance documents. Some prefer to contain the requirements in a single document. Some prefer to split up the guidance by topic and others by building a risk profile. There is probably reluctance in most, if not all, countries to entirely start from scratch. The risk-averse nature behind fire safety advocates caution. That hesitancy will lead to progressive and step-by-step adjustments to their fire safety codes. It is clear that the more modern approach behind performance setting fire engineered solutions is the only likely route to greater levels of uniformity

Question 2: Describe the key criteria for means of escape?

The Chinese Code bases requirements on building height and whether the building is sprinkler protected. There are no limiting criteria on vertical travel distance. Furthermore, there are no special requirements for the evacuation of mobility impaired persons. Where a performance-based design is proposed, then ASET/RSET criteria are typically used. Note that Hong Kong did have extensive requirements before 1997 for means of escape as specified in their Code of Practice for provision of means of escape in case of fire[31]. The Code incorporated a series of tables for the critical dimensions, including travel distances that are not dissimilar from British Standards. Some differences did exist. For instance, exit routes should have a clear head height of more than 2m. Buildings less than six storeys are permitted to have a single escape staircase.

Indian Standard IS1644 bases the requirements for means of escape on the building profile. In general, every floor of a building should have a minimum of two separate exit routes. The Indian Standard also determines the total width of exit routes based upon occupancy numbers and the type of exit route (stairway, ramp, or door). A standard unit exit width is given as 50cm. Occupancy numbers per exit width can vary from 25 for residential or educational buildings to 75 for most categories of building profile. Note also that additional factors may be used to allow up to 100% increases in occupancy numbers. The typical travel distance to an exit is 22.5m, although this can increase to 30m or even 45m for industrial buildings. Given the age of the standards, issues such as RSET/ASET are not included. However, it is understood that performance-based approaches are increasingly being adopted in the country.

Iran uses tables to determine the means of escape criteria, including occupancy type and numbers, as well as if the building is sprinkler protected. For example, for each floor of a building, a single escape route is acceptable for up to 500 occupants, two escape routes for occupancy numbers between 501 to 1,000, and four escape routes for occupancy numbers greater than 1,000. For other building types such as residential units, industrial buildings, car parks, etc., a single route is acceptable. Escape route widths are based upon a table but should typically be a minimum of 1,100mm, although in some identified cases, 900mm is permitted. For large projects, RSET / ASET analysis may be used, but this is rare

New Zealand permits a single means of escape for up to 500 people, based on the floor or total area. As an example of travel distances for sleeping risks, dead-end "open path" travel distances (where exit can only be assumed in one direction) should be between 25m to 50m. The distances are determined by whether there are fire detection and sprinkler systems in place. It also states that total "open path" travel distances (i.e., to a place of safety) can vary between 60m to 120m. For offices, the maximum travel distances expand to 75m and 150m, respectively. Minimum escape route widths are stated as 850mm for horizontal travel and 1,000mm for vertical travel (for offices). Some variations to this are deemed acceptable under certain specified conditions.

The Polish technical regulations for buildings generally measure escape distance to an enclosed staircase fitted with fire doors or to a final place of safety. For the highest risk category (explosive zone) or for buildings with higher occupancy densities or used as hotels, the one-way travel distance is only 10m and increases to 40m where two-way escape is possible. Standard public buildings increase these two figures respectively to 30m and 60m. For warehouses with relatively low fire loading or for many types of dwellings, the values increase again to 60m and 100m, respectively. These figures may be increased by 50% when either sprinkler systems or smoke control systems are installed. When both system types are used, a 100% increase is permitted. Where smoke control systems are specified, it is obligatory to prove their performance by an engineering assessment.

The United Kingdom requirements derive from their Building Regulations, and specifically Approved Document B of the Regulations. BS9999, however, allows for some additional variation based upon risk profile. BS9999 provides a limiting factor for single escape routes of a maximum occupant capacity of 60 in a room, tier, or storey, and the travel distance limit for travel in one direction only is not exceeded. For situations where between 60 and 600 occupants are on a floor, two separate escape routes should be provided. For occupancies over 600, three separate routes are required.

Travel distance is also based upon risk profile. The maximum allowed for low-risk buildings is 90m for two-way travel and 30m for one-way-only travel. For offices, these figures are 75m and 26m respectively. In addition, allowances are given for sprinkler installations. Dimensioning of escape routes is also based upon risk profile.

In the case of the United States, the means of egress will vary slightly due to the building profile. Travel distances are measured from any point on a floor to the protected escape route (e.g., stairway). A single exit is allowed for areas with a total occupant load of fewer than 100 persons, provided that the exit shall discharge directly to the outside, and the total distance of travel from any point, including travel within the exit, does not exceed 100ft (approx. 30 m). This also applies to buildings with less than three floors, with some exceptions. In other cases, duplicate exits are required where travel distances to an exit do not exceed 200ft (61m) and 300ft (91m) in business occupancies protected throughout by an automatic sprinkler system. Dead-end corridors must not exceed 50ft (15m). Some of the acceptable dimensioning of means of escape are also based upon occupancy numbers and building profile/height.

It is clear that there is a relatively high degree of consensus when it comes to means of escape criteria. Thus, even with some variations in aspects such as travel distance, it is not inconceivable that a global prescriptive set of rules for means of escape could be acceptable to all.

Question 3: What are the essential requirements for fire compartmentation?

The Chinese Code bases structural and internal fire compartmentation on building height and use. Internally, a fire-resistance rating of up to 180 minutes is specified. Protected staircases are afforded 120 minutes fire separation from risk areas. Ultra-high-rise buildings such as Shanghai Tower[32] must have a minimum of 180 minutes fire resistance against fire-induced progressive collapse.

Indian Code IS1642 bases fire resistance requirements on four building "Types". The standard quotes a minimum fire resistance for any *Type* for separation of exit way corridors as 60 minutes. Shafts and stairways for all building types are required to meet a minimum of 120 minutes fire resistance. For Type 1 buildings (highest risk) – load-bearing elements of exterior walls and firewalls, the quoted fire resistance is 240 minutes. The Standard also provides a range of building material options to meet the requirements. For buildings higher than 15m, all floors should be compartmented with a maximum fire zone of 750m² (or 500m² for profiles such as shopping centres). Where a sprinkler system is installed, then this fire zone size can be increased by 50%. The maximum distance between compartment walls is specified as 40m.

The Iran National Building Regulation bases compartmentation requirements on ten occupancy types, including residential, educational, health care, industrial, offices, and warehouses. Then, specific to each type, varying requirements are given for different building elements, including load-bearing members, exterior walls, separation walls between compartments, staircases, corridors, etc. The minimum rating is 60 minutes fire resistance, and the maximum rating is 180 minutes. The installation of a sprinkler system also has an impact. For example, load-bearing elements for non-sprinklered buildings are specified at 180 minutes, reducing this figure to 60 minutes for buildings with a sprinkler system. For separation walls between an industrial occupancy and public occupancy (such as ceremony halls, restaurants, waiting halls), the fire resistance should be set at 180 minutes.

The New Zealand criteria for offices and commercial spaces (life rating) sets the minimum fire resistance as 60 minutes and 120 minutes for property protection. Furthermore, the latter value may be increased to 180 minutes for factors such as storage height and location of the boundary. Where sprinkler systems are fitted, then these figures can be halved. The minimum fire-resistance rating for residential apartments is given as 30 minutes. For high-risk buildings such as warehouses, 240 minutes is the maximum level of fire resistance required.

The Polish Technical Regulations base the construction elements of fire resistance and compartmentation on building height and profile. They divide buildings into five distinct classes (A to E) based upon a combination of building profile and height. The class may be decreased when either sprinkler systems or smoke control systems are used. These classes can then be used to define the fire resistance properties of building elements and fire separations. For instance, the highest, class A, recommends that the main construction elements of the building are rated at 240 minutes.

In contrast, category E does not have a fire-resistance requirement other than for the separation between two fire zones, where the fire resistance required can be a minimum of 15 minutes. Furthermore, maximum fire zones are limited from 2,000m^2 to 10,000m^2 for standard public buildings and 20,000m^2 in warehouses. The fire zones may be increased by 50% when either sprinkler systems or smoke control systems are used.

United Kingdom's BS9999 once again uses risk profiling and the option of sprinkler system protection to adjust requirements for both structural and internal fire resistance. Typically, internal fire separations between rooms and corridors should have a 30-minute rating. A 60-minute fire resistance is generally required to separate firefighting shafts from lobbies, with a total of 120 minutes fire separation between firefighting shafts and risk areas of the building.

The United States Code NFPA 101 provides a table (Table 6.1.14.4.1(a) - Required Separation of Occupancies) covering the fire resistance separations between occupancies based on profile. For example, the fire separation between industrial units and other profile types may be up to 180 minutes. Typically, primary fire separations should be rated at 120 minutes. Internally, the code requires exit routes to incorporate fire separations with a minimum of a 60-minute fire-resistance rating, where the exit connects three or fewer floor levels. Otherwise, the separation must have a 120-minute fire resistance. Half-hour fire-resisting separations are recognised for lower risks.

For what is probably the oldest form of fire protection – passive fire protection, it is somewhat surprising to identify such a variation in ratings. Possibly these variations may reduce as more pan-national standards are introduced and accepted. In this case, I am thinking specifically of the work to standardise fire safety for building construction ongoing in Europe.

A European classification system for reaction to fire recognises that a single fire resistance rating value is insufficient. EN 13501-1[33] provides further information. Aspects such as load-bearing capacity, integrity, insulation, and so on can all be tested separately. Consequently, a European rating system can provide much more information. The following fire resistance parameters and properties can be set individually and combined into a single rating:

R	Load bearing capacity
E	Integrity
I	Insulation
W	Radiation
M	Mechanical resistance
C	Self closure
S	Smoke leakage
K	Fire protection (coverings)

For example, a fire compartment requiring a load-bearing capacity, integrity, and insulation rating of 60 minutes, will be given as REI 60. A 30-minute fire door with smoke seals will be identified as E30S.

Question 4: What are the primary factors for fire detection and alarm systems?

The Chinese Code applies categories of fire detection systems based upon the requirements for the building. Siting and spacing requirements relate to the type of detectors used. In the event of an alarm, the majority of buildings use a simultaneous evacuation strategy.

In the case of India, the relevant standard is IS 2189[34]. On review of this code, it closely covers the requirement of British Standard BS 5839-1. However, the categories of system, as included within the BS, are not utilised.

Iran can make use of either the UK BS 5839-1 or US Code NFPA 72. The categories listed in BS 5839-1 are often specified.

New Zealand uses NZ Standard 4512[35] for fire detection and alarm systems. This Code categorises systems into various "Types". Note that these classifications are contained in an appendix and are provided as supporting information. The types themselves list multiple options, from simple manual systems, with or without automatic signalling to a remote receiving centre, to the inclusion of heat or heat/smoke detectors. Sprinkler systems are also classified as heat detection systems. In terms of detector siting and spacing, point-type smoke detectors should be placed so that there is a maximum of 10m between detectors and no point in a room is more than 6m from a detector (and 5m from a wall). The criteria for heat detectors are that the maximum distance between devices is 6m, and this can be increased to 9m in corridors. They should be located within 3m of a wall (4.5m in corridors) and at a density of 30m^2 per device. Voice alarm systems are becoming more commonly specified in NZ.

Polish requirements for the use of fire detection and alarm systems are regulated by the ordinance of the Ministry of Interior and Administration. The criteria for where fire detection and alarm systems are required are based upon the building profile and fire zone size. As with many other areas, component standards make use of a range of European standards. Fire detector spacing sets the maximum distance from any point to the nearest heat detector at 5m and, for smoke detectors, at 6m to 7.5m.

The United Kingdom's predominant standard for fire detection and alarm systems for non-residential buildings is British Standard BS 5839-1[36]. It is also specified in many other countries. The BS uses a set of categories to define the extent of detection as well as the objective(s) – manual only system (M), life safety (L1 to L5), and property protection (P1 and P2). The spacing parameter for smoke detectors is set at 7.5m from any point in the room to the device. This is reduced to 5.3m for heat detectors. Note adjustments to these values are allowed subject to locational conditions. Alarm sound levels are typically stated as 65dbA or 5dBA above ambient and 75dBA at bed heads for sleeping risks.

The United States' primary code for fire detection and alarm systems is NFPA 72[37]. This Code covers all aspects of detection and signalling system, including requirements for control equipment, cabling, alarm sounders, and siting and spacing of fire detectors. For example, NFPA 72 requires that smoke detectors spacing on ceilings is 30ft (9m) point-to-point and 15ft (4.5m) to walls. For heat detectors, the spacing considers height. For a 10ft ceiling, the spacing should be no more than 12ft (3.6m). Audible signal appliances intended for operation in the public mode should have a sound level of not less than 75dBA at 10ft (3m). Sound levels for sleeping areas should be a minimum of 70dBA or 15dBA above the average ambient sound level.

Question 5: Requirements for firefighter response?

In China, there is no specification for the expected time of response by firefighters. The one provision is that external fire hydrants should are provided within a 150m radius of the building.

The Indian Standards do not go into detail regarding firefighter attendance times. Indian Standard IS3844[38] covers internal fire hydrants and hose reels on premises. The requirement for firefighting provisions is based upon building height and use. Buildings over 15m should incorporate provisions for wet riser systems, with the number of risers required based upon building type and floor area. Typically one riser is required per 500m^2 to 1,000m^2. The tank storage requirements increase with building height.

In Iran, the actual maximum attendance time by the emergency services in Tehran is set at 4.5 minutes. The goal is to reach 2.5 minutes. Fire hydrants should be provided within 75m of critical building locations, such as high-density residential areas and commercial and industrial zones. In regions with average population density and low-risk commercial buildings, this is set at 100m and increases to 150m for low-risk areas with a low population density. For high-risk facilities such as warehouses, two to three hydrants are deemed necessary. Wet and dry risers in buildings should meet US Code NFPA 14[39].

New Zealand recommends that the maximum distance from the point at which the fire appliance can park to the riser inlet into the building is 75m unless a sprinkler system is fitted. Buildings higher than 10m should have firefighter lift control features. Attendance time depends upon regionality and whether the building is in urban or rural locations.

Polish regulations typically assume that the attendance time of the fire and rescue services will be within 15 minutes. In some cases, the services ask for a special declaration over the expected time to attend. The requirement for firefighter shafts is based upon building height and profile. For commercial and office buildings, for instance, shafts are required for those higher than 25m. For residential blocks, this increases to 50m. Note that dry risers are not a specific requirement for inclusion within firefighter shafts, although a form of smoke control is.

The United Kingdom offers a range of requirements for firefighter facilities. Buildings higher than 18m should typically be fitted with dry riser systems and firefighting shafts. Hydrants should be provided within 90m of dry fire main inlets on a route suitable for laying hose. For buildings without fire mains, hydrants should be provided within 90m of an entry point to the building and not more than 90m apart. The UK Government regularly publishes achieved attendance times.

NFPA Code 101 covers the US requirements in buildings for appropriate controls, including control of elevators, smoke extract, ventilation systems, etc. In addition, there is a code to cover firefighting facilities. NFPA 1[40] provides fire and life safety guidance for the public and first responders. It also includes property protection in its scope with a comprehensive, integrated approach to fire code regulation and hazard management.

For example, it requires that all fire hydrants' aggregate fire flow capacity within 1000ft (305m) of a building cannot be less than the required fire flow. A table helps determine the number of fire hydrants required and their distance from the building. A further code has been produced, NFPA 1710[41], to cover response times, and gives a measurement system from initial alarm to actual firefighting. Criteria include:

- Alarm answering time: 15 seconds for 95% of calls; 40 seconds for 99% of calls;

- Alarm processing time: 64 seconds for 90% of calls; 106 seconds for 95% of calls;

- Turnout time: 60 seconds for EMS responses; 80 seconds for fire response;

- First Engine arriving at the scene: 240 seconds for 90% of responses with a minimum staffing of 4 persons;

- Second Engine arrives at the scene: 360 seconds for 90% of responses with a minimum staffing of 4 persons.

Firefighter criteria, as expected, will be based mainly upon the specific setup criteria of the fire services and their particular requirements. But this is a hurdle that could be overcome.

Summary of findings

The idea for a global comparison of standards seemed to be a sensible exercise for my research. However, I quickly realised that a simple like for like analysis proved more difficult than expected. This was mainly due to national code styles and my format for questioning. Nevertheless, I can make some early conclusions:

1. The criteria for means of escape did show some variations. However, the differences were relatively minor, and, with goodwill, a global set of means of escape criteria could be realistically achieved.

2. The application of fire compartmentation showed a surprising degree of variation, given that construction styles and methodologies should be interchangeable anywhere in the modern world.

3. Agreements for a global standard for fire detection systems should be straight forward although they could be influenced by a preference for either the UK or US approach.

4. The basic facilities for firefighting are similar, although some of the detailed requirements may vary.

5. Most countries now acknowledge the benefits of "trade-offs," such as a relaxation in requirements if, for example, fire sprinkler systems are used.

No doubt, a more in-depth examination could be helpful, with an increased number of questions and a much more comprehensive analysis of countries, or even regions within countries. My gut feel is that a fully global approach is achievable but will probably rely on the willingness of all involved. In particular, the UK and US would need to be more open to understanding the strengths and weaknesses behind their national approaches.

I also believe that the embracing and complete acceptance of performance-based fire engineering will be the catalyst in the global alignment of fire safety standardisation.

5 The interdependency of fire engineering with other core disciplines

Different engineering disciplines share common components and dynamics.

Fire engineering emerged in the early 20th century as a distinct discipline, separate from other engineering groups, in response to new fire problems posed by the growth of factories and the special fire hazards they introduced[42]. Due credit for the early advancement of fire safety and protection must be given to the insurance industry and the City of London insurance market in particular. But, of course, I may be slightly biased given that I started my career in a City of London historic insurance organisation – the Fire Offices' Committee (see below).

The Sun Fire Office[43] innovated fire safety for buildings by issuing fire marks (metal plaques identified with the insurer's name) from 1710. These marks were used to help identify those insured premises so that the insurer's fire brigade would attend to the fire. Formalising such services helped improve the practice and consistency of firefighting, and the idea was quickly taken on elsewhere.

The Fire Offices Committee (FOC)[44] was formed in 1868 by British fire insurers, primarily for insurance tariff setting for their insured risks. The FOC were early pioneers in the advancement of knowledge of fire safety and control measures to limit their financial exposure.

They produced early codes for elements of passive and active fire protection systems. They also gradually introduced schemes for approving fire protection from firewalls and linings to sprinkler and detection systems. The scheme involved a technical assessment of components of a system and the interaction of the components as part of a system — this usually involved testing. At one time, the UK Insurers even had their test laboratory – the Fire Insurers Research and Testing Organisation (FIRTO), located in Borehamwood, UK. In 1953, a European insurers organisation was set up to help develop European rather than national interests. The organisation was known as the Le Comité Européen des Assurances (CEA) until 2012, when it changed its name to "Insurance Europe" and is currently based in Brussels. One of their objectives was, and is, to standardise a Europe-wide response to insurance issues, including fire safety, with a focus on property protection.

The USA has its agencies representing insurers' interests – Factory Mutual (FM) and Underwriters' Laboratories (UL). The origins of FM date back to the 1830s when a mill owner formed his own mutual insurance company with other factory owners. Today, FM[45] is a globally recognised property and business insurance company incorporating risk professionals who evaluate and advise on risks and provide test facilities. UL[46] was also formed in the latter part of the nineteenth century via a proposal to create a test laboratory for the benefit of US insurance underwriters. They state that one of their objectives is to promote safe, secure, and sustainable living and working environments for people by applying science, hazard-based safety engineering, and data insight.

Insurers have played a vital role across nations in helping develop the science and engineering behind fire engineering. The work of insurance organisations influenced many of the fire safety codes mentioned in this book.

Fire engineering, as an area of study, has only been acknowledged in relatively recent times. It is claimed that the first formal degree course was introduced in the United States in1903[47] by the Armour Institute of Technology (later becoming part of the Illinois Institute of Technology). At the same time, the College introduced a core discipline of chemical engineering.

The advent of formalised education in the subject was in the 1970s and onwards. The University of Edinburgh is well recognised as a focal point for study and attracted eminent names. Fire engineering icons such as Professor David Rasbash and Professor Dougal Drysdale (probably names familiar to the more mature readership) were both involved with the University.

From this point forward, the subject, the science, and engineering behind it have continuously developed, with graduate and post-graduate courses now available on an international basis.

The primary representative bodies for the profession are the Institution of Fire Engineers (IFE) and the Society of Fire Protection Engineers (SFPE). The IFE[48] was founded in the United Kingdom in 1918 and represented the interests of those practitioners from graduate to chartered engineer level. It is now recognised as an international institution with members from most parts of the world.

The SFPE[49] was established in 1950 in the United States and was incorporated as an independent organisation in 1971. It states that it is the professional society representing those practicing the field of fire protection engineering.

The story given in this chapter so far reflects that fire engineering is largely well represented, derives from robust bones, and can look forward to future generations of fire professionals. Nonetheless, those not in the profession may fail to understand how broad the subject is and how it draws on the knowledge and experience of many other areas of engineering. This can be best illustrated by the diagram given in Figure 3.

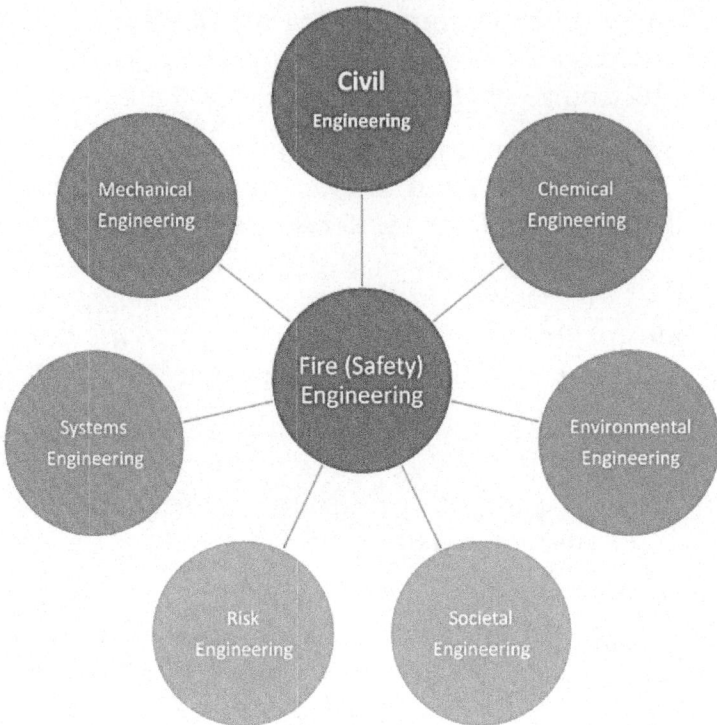

Figure 3: The interdependency of fire engineering with other disciplines

Civil Engineering[50] is neatly described as everything to do with the built environment. As well as the discipline behind the construction of buildings, civil engineers also design roads, railways, and infrastructures such as water and gas pipelines. A subset of this discipline is structural engineering, described as the study of the design of structures and non-structural elements that bear a load. Of critical importance is in understanding how structures respond to fire exposure. Consequently, this is of direct relevance to fire engineering. Furthermore, civil engineering is the primary discipline to increasingly forward the use of BIM (Building Information Modelling). The issues of BIM and fire engineering are covered later in this book.

Societal engineering[51] is described as the study of the creation and influence of human societies. It is a field of social science dealing with those social dynamics that operate on a large enough scale to affect entire populations. It uses both philosophical and practical approaches to improve our life. The aim of fire engineering is also to improve our life by making buildings safer. The psychology of human behaviour in a fire is a deciding factor in designing means of escape. Note that there is a separate scope of social engineering related to internet security which is not relevant to this discussion.

Systems (and industrial) engineering[52] is concerned with designing, improving, and installing integrated systems of people, materials, information, equipment, and energy. It draws upon specialized knowledge and skill in the mathematical, physical, and social sciences and the principles and methods of engineering analysis and design to specify, predict, and evaluate the results obtained from such systems.

Fire engineering is an excellent example of a systems engineering approach. The whole purpose of a fire strategy is to consider several factors covering the building, processes, and occupancy profiles and provide an integrated set of complementary systems (fire detection, control, suppression, etc.) to satisfy a range of fire scenarios.

There are two variants in the understanding of what **environmental engineering** is. The most simplistic[53] is to control a working environment for process requirements or people's comfort, i.e., heating and ventilation systems. A US[54] description of the discipline of environmental engineering defines it as the application of engineering principles to improve and maintain the environment for the protection of human health, for the protection of nature's beneficial ecosystems, and environment-related enhancement of the quality of human life. Both understandings are relevant to fire engineering. The internal environment can play a part in fire ignition and growth and control measures such as the maintenance of tenable conditions for escape. The broader context of the impact of fire and firefighting on the environment is similarly highly relevant to future applications of fire engineering.

Risk engineering[55] is the application of engineering skills and methodologies to the management of risk. It involves hazard identification, risk analysis, risk evaluation, and risk treatment. This subject is directly relevant to fire risk analysis and scenario determination, as described later in this book.

Chemical engineering[56] is a discipline influencing many other areas of technology. Chemical engineers conceive and design processes to produce, transform and transport materials from experimentation in the laboratory to full-scale production. The scientific principles behind chemical engineering are fundamental to the understanding of fire ignition and growth. Furthermore, the engineering discipline is relevant to fire control, particularly when it comes to fire suppression technologies. The influences of chemistry and chemical engineering are considered later in this chapter.

Mechanical engineering[57] is a branch of engineering concerned primarily with the industrial application of mechanics and the production and use of tools, machinery, and products. This is the discipline that applies physics, mathematics, and materials science. It is thought to be one of the oldest and broadest of all engineering disciplines. In many ways, mechanical engineering is like chemical engineering in that it helps to understand the forces behind fire growth and spread. It is also relevant to control means such as the use of mechanical smoke control systems.

There are, of course, a few secondary disciplines that will impact fire engineering albeit, not directly. These include electrical/electronic engineering, computer engineering, and materials science and engineering.

Each of the primary disciplines is relevant to one or more elements of every fire strategy. This can be explained by considering the fire strategy value grid (taken from British Standards Specification PAS 911) and each of the eight elements given (see Figure 4).

Table 1 provides a list of how I believe each of the primary engineering disciplines can be applied to each of the eight fire safety factors. What can be revealed is that both chemical and risk engineering play a dominant role in all aspects of a fire strategy. Risk engineering is considered further in the next chapter.

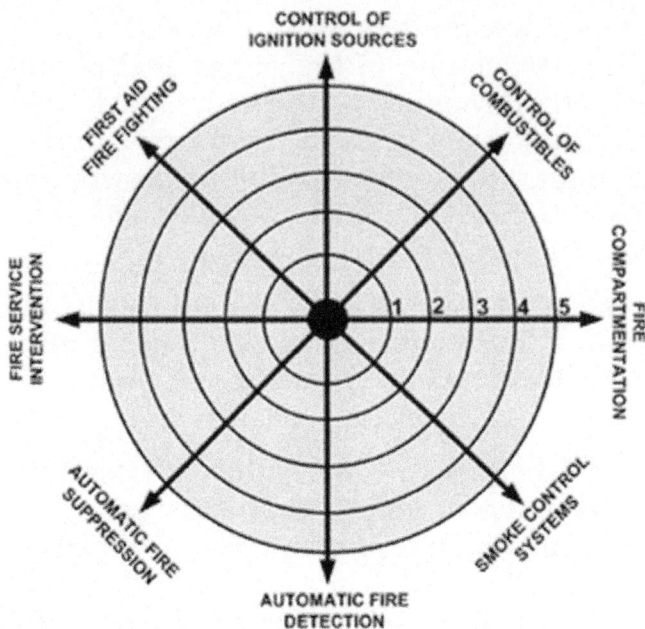

Figure 4: Fire strategy value grid (BS PAS 911)

Table 1: Engineering discipline interplay with a fire strategy

Fire strategy input	Primary engineering discipline
Control of ignition sources	Chemical, Risk
Control of combustibles	Chemical, Risk, Environmental
Fire compartmentation	Chemical, Risk, Mechanical, Civil, Systems
Smoke control systems	Chemical, Mechanical, Systems, Environmental, Risk
Automatic fire detection	Chemical, Mechanical, Systems, Environmental, Risk
Automatic fire suppression	Chemical, Mechanical, Systems, Environmental, Risk
Fire service intervention	Chemical, Mechanical, Systems, Environmental, Risk, Societal
First aid fire fighting	Chemical, Environmental, Risk, Societal

Chemical engineering combines the study of a range of engineering and scientific principles, including chemistry, physics, mathematics, biology, and economics. It is an applied study of how gases, liquids, and solids react, absorb, distill, spread, and transform.

Although chemical engineering conjures up pictures of large chemical processing plants, its primary foundation is fundamental to both the study of fire and how the effects of fire can be mitigated/controlled. Fire is a chemical reaction. The ignition of fire requires a basic chemical process.

The growth of fire around a building follows some standard principles of chemistry. All these elements are captured in the subject of chemical engineering.

The creation of fire is described as an "exothermic chemical process of combustion", releasing heat, light, and various reaction products. At a certain point in the combustion reaction, called the ignition point, flames are produced. Flames consist primarily of carbon dioxide, water vapour, oxygen, and nitrogen.

An exothermic reaction can be demonstrated by a time/energy graph (Figure 5).

Note that the system's internal energy initially increases, followed by a decrease as the internal energy is

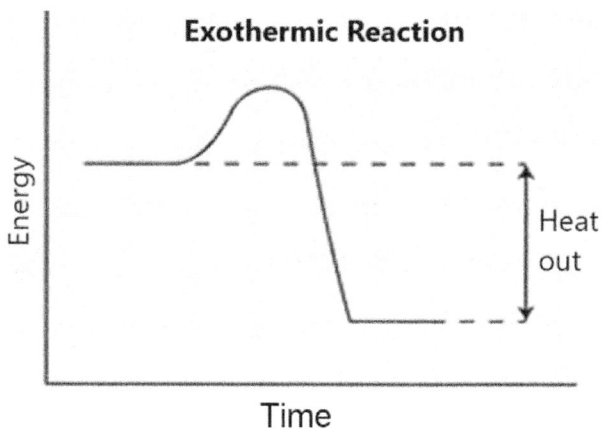

Figure 5: The exothermic time/energy graph

converted to mostly heat.

An understanding of transport phenomena can explain the main principles and processes of fire growth. This topic encompasses the analysis of the exchange of mass, energy, and momentum within systems. The parameters can be assessed using mathematical equations, which consider two- and three-dimensional flows. The analysis uses the assumptions behind energy conservation, i.e., the phenomena are considered individually, and the sum of their contributions will equal zero. This principle helps calculate many relevant quantities, such as the velocity profile of a fluid flowing through a rigid volume.

Transport phenomena allow for an understanding of how fire and smoke growth is limited or exacerbated by heat and mass transfer conditions. Furthermore, the calculus behind transport phenomena can be exploited to assess how active and passive fire protection can provide suitable mitigating conditions. There are three primary circumstances for heat transfer:

Thermal conduction: Heat transfer occurs at a molecular level when the heat energy from a fire moves as thermal, kinetic, and through a medium (primarily solid, but also liquid or gas). Typically, conductive heat transfer flows from areas of high temperature to areas of low temperature.

Thermal convection: Convective heat transfer is when heat travels from one place to another by the movement of fluids (liquids and gases, including air). Within an enclosure, thermal convection can be modelled as smoke is carried along with the convection currents.

Thermal radiation: Thermal radiation is heat transfer via electromagnetic wave energy. It allows us to see and feel the direct influence of fire. We can even tell the type and constituent parts of the burning materials via the colour of the burning properties

It is the principles of thermodynamics that will dictate how an ignition source will combine with combustible materials. For example, using a blow torch within proximity of combustible materials can cause combustion via convection and radiation. Even a domestic blow torch used for culinary purposes can provide flames above 1,300^{0}C. The total radiative emissive power will be considerable for even a small-sized flame from such a tool. This is because the emissive power is proportional to the fourth power of the temperature. Several fires where construction activities, particularly hot works, have led to the destruction of building sites are the result. By understanding the processes behind both the initial combustion phase and then by the methods of thermal spread, we can gain a more fundamental understanding of fire scenarios. Table 2 provides a few examples.

The science of fire spread will be considered later in this book to determine a possibly more effective methodology to assess the risk of fire growth - and how to control it.

Even though fire engineering is aligned with chemical and risk engineering, many believe that civil engineering is the closest discipline, particularly structural engineering. The relationship is understood within both professions, and one is often seen as a core element of the other. A quick search on the internet will reveal combined fire and civil engineering graduate courses.

Table 2: Means of smoke and fire spread with control measures

The main method of fire spread	Scenarios
Convection	Smoke and flame spread, say, from an unventilated room into corridors is primarily convective and can be controlled via passive fire protection. Alternatively, passive or active smoke control can be used to divert smoke and flammable gases. An alternative is to limit the fire by suppressing it at the source, using a sprinkler system. Another example is smoke spread within an atrium. Control measures could include ventilation at high level, fire curtains to prevent smoke egress into upper open levels, and suitable incoming air supplies to provide tenable conditions for evacuation.
Conduction	A fire in a cable tunnel can travel via the copper conductors of the cable. Control measures could include early detection of the rate of heat rise, removing the source of electrical power, and possibly using local application suppression systems.
Radiation	The primary means by which fire spreads from one building to the next is largely radiative. Many national fire codes highlight this as a specific issue. One control method is to limit the percentage of the unprotected area of the external fascia of buildings. This percentage is the amount of, for instance, glazing used in a brick-built building. The greater the levels of the unprotected area will mean more significant levels of heat radiation from a fire travelling between neighbouring buildings.

Structural engineering covers a range of subjects concerning construction techniques and the materials used. A primary component is in the understanding of shear stresses of both horizontal and vertical components, deflection in beams, truss, and frame structures, as well as column and beam buckling. The performance of materials such as steel, concrete, and timber is an equally important consideration.

A notable example of how fire engineering influences structural engineering is understanding tensile membrane action (TMA). Designs for steel frame buildings had focused on controlling the temperature of individual steel members rather than considering the structural performance of a building, holistically, in the event of a fire. All structural steel columns and supports should be contained within fire-resisting cladding or other materials that protect a fire. Wang[58] highlights that the ability of a reinforced concrete slab to "bridge" over damaged loadbearing steel beams is the result of TMA. In such situations, floor slabs can withstand loads many times higher than the design strength of the floor at small deflections. TMA has significant implications for the necessity of fire resistance and fire protection of steel-framed buildings. Bailey[59] had researched a fire engineering design for multiple-storey, steel frame buildings that eliminated the need for fire protection for up to 40% of steel floor beams, with significant savings.

Transport and Tunnel Engineering is also a specialist civil engineering discipline. It covers road, rail, bridge, and airport design and the related infrastructure, such as railway stations, airport termini, etc. In some cases, the specific considerations of the impact on fire have most influenced the approach to civil engineering design. One such UK example was the introduction of the Fire Precautions (Sub-surface Railway Stations) Regulations in 1989. This Act was introduced following the Kings' Cross underground fire in London in 1987, where 31 persons lost their life due to a fire engulfing a sub-surface concourse. This piece of legislation radically updated design criteria for underground railway stations in the UK.

The construction of tunnels has been similarly influenced by the need to ensure appropriate fire safety and protection levels. There have been numerous fire incidents in tunnels over recent decades. One notable example often used to illustrate the critical issues is that of the Mont Blanc tunnel. This Tunnel is a highway tunnel under Mont Blanc Mountain in the Alps. It links Chamonix, Haute-Savoie, France with Courmayeur, Aosta Valley, Italy. This tunnel experienced a significant fire in 1999[60]. A Belgian truck carrying margarine and flour caught fire and stopped almost 7 km into the tunnel. The fire quickly spread to the vehicles behind the truck for over 1 km. The intense heat and smoke filled the entire tunnel section, preventing emergency rescue and firefighting operations. The fire burned for two days and reached temperatures of 1000°C, killing 39 people. Most drivers stayed in or near their vehicles. Those who tried to escape collapsed due to smoke inhalation. The CO content in the smoke was understood to rise quickly to over 150 ppm within minutes. The fire led to several changes in tunnel design. Tunnel lining materials were upgraded. The frequency and availability of "safe areas" were also improved. Given the increase in tunnel building, particularly in Scandinavia, specialist fire safety in tunnel conferences have become extremely popular, with new concepts continually being offered to advance tunnel construction criteria.

Geotechnical (Soil) Engineering recognises that all structures such as buildings, roads, bridges, airports are supported on the ground. The ground characteristics play a vital role in the stability of the building above or below. This subject covers the characterisation and classification of soils and testing procedures. One determinant is in the understanding of how the ground may be influenced by water. In a huge fire, thousands of litres of water or foam will be used to fight it. Firefighting water run-off is habitually the result and could completely saturate the soil. A thought for the fire strategy – possibly?.

The message of this chapter is to highlight how complex the subject of fire engineering is and how there is a degree of dependency on a range of other engineering disciplines. Those who work outside the field may not realise this. On many occasions, a fire engineer is expected to have an in-depth knowledge of every aspect of the subject. In reality, fire engineers tend to have specific areas of expertise with possibly a passing knowledge of other factors.

For instance, Gillian may have incredible levels of expertise in computational fluid dynamics and modelling fire behaviour, but she may not have the same depth of knowledge in the efficacy of fire sprinkler systems. A cause-and-effect logic diagram commonly used by Mohammad, a fire detection and alarm specialist, may prove baffling for Professor Armand – a lecturer in fire sciences. John, an expert in structural fire engineering, may not appreciate the finer points of fire door construction.

And let us broaden this still further given that somebody knowledgeable of smoke control standards in the UK may know very little of smoke control requirements in the US.

If we multiply the sheer number of specialisms that make up fire engineering with the national permutations, then the result is mind-boggling. Does this mean that the search for a single global methodology stops here?

Paul Bryant

6 Sustainability and BIM – two issues that will increasingly impact fire strategies

Progress is a driver for amended thinking.

The previous chapter highlighted the complexity of fire engineering and the range of specialisms within the single embracing term. Nevertheless, fire engineers will need to be aware of what is going on in the wider world when developing fire strategies. Two issues are increasing in prominence, and it would be remiss of me not to address them within a book promoting a new way of thinking. The areas of growth I believe will impact the most in coming years are sustainability and the technology of Building Information Modelling - or BIM as it is more commonly referred.

Sustainability and fire safety

Fire engineers aim to ensure that the building they are involved with is *firesafe* -protecting the persons within the building. A noble cause – although is it a "sustainable" cause? Collins English Dictionary defincs sustainability as the ability (for a scenario) to be maintained at a steady level without exhausting natural resources or causing severe ecological damage.

Given that damage control is also a primary objective of fire safety and protection, it could be argued that the goals of both the sustainability and fire safety industries are aligned. But this is not always the case. Some of the methods used to fight fires are harmful to the environment. For example, Halons (halogenated hydrocarbons), once prominently used to extinguish fires, were found to be highly hazardous to the ozone layer and were banned for general use in the 1990s. In addition, run-off water from the action of firefighting typically can contain chemical bi-products of combustion, leading to contaminated land, rivers, etc.

The extensive requirement for fire compartmentation in building fire strategies may be seen as a mismatch with sustainable building design, favouring more open building layouts.

If sustainability is to become a natural feature of the built environment, then the methods used to apply fire safety need to incorporate elements that consider the environment. Let's consider how fire strategies are both prepared and evaluated. Troisi et al.[61] refer to enforcement authorities as one agent in improving a systemic approach to safety systems, with a shift of perspective from single bodies involved with safety performance to interdependence with all agents.

As previously concluded, despite the advancement of fire-engineered solutions, most buildings are still required to meet with the prescriptive national building regulations. The whole focus is on rigid designs to help protect building occupants against the ravages of a fire, with less concern for other aspects such as the internal and external environment of the building.

The Society of Fire Protection Engineers published an article[62] that stated that performance-based fire engineering design will become increasingly important in ensuring sustainable building design. The article recognised the conflict between fire safety prescription and sustainability issues. It suggests that to reduce a building's resource use quantifiably, fire engineers need to consider the resource demands of products necessary to achieve the required fire safety performance. Likewise, the fire protection industry needs to recognise new materials and products and adapt the fire safety solutions accordingly.

A manufacturer of fire-resisting glazing[63] highlighted that fire could significantly impact the sustainability of communities and the built environment since a building that burns down is fundamentally unsustainable. They point out that fire protection of buildings has an increasing role on the sustainability agenda. Yet, most fire standards and codes rarely highlight the issue of sustainability. It is only by promoting an innovative approach can the subject and objectives of sustainability be expressly incorporated.

Figure 6 illustrates a model depicting how sustainability relates to the critical factors of society (social equity), economics, and the environment. The balancing of all three elements will enable true sustainability to be delivered. But could this relate to the objectives of fire engineering? Social factors closely relate to the ethical need to protect life, and this is provided by national fire safety regulation as a fundamental part of all building legislation.

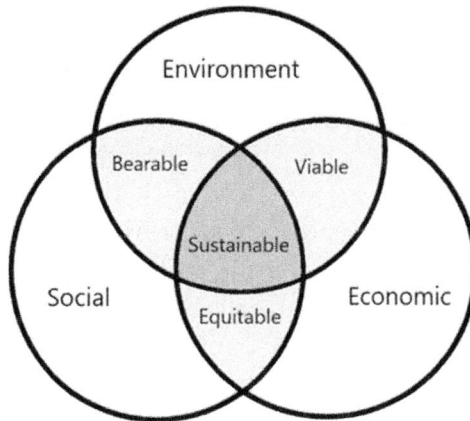

Figure 7: A sustainability model

The environmental issue is less touched upon by prescriptive fire standards but can be a crucial consideration and a performance objective for modern fire safety engineering. Economics is a factor that is relevant in both approaches. The added flexibility of fire engineering will deliver benefits by avoiding what I have described as "overprotection" and applying systems as specifically required.

The obvious conclusion is that adopting performance-based techniques for the specification of fire safety and protection will assist in developing "sustainable buildings". Some issues may prevent the methods from being globally approved, as discussed in this book. Nevertheless, the drive towards "green" buildings and core sustainability in community design continues. The fire industry will need to adapt.

Building information modelling (BIM) and fire safety

Building information modelling, or more commonly referred to in its abbreviated form – BIM, is widely regarded as the next inevitable step in building design and construction. Some would go so far as stating that it is the *currently* preferred method and that those who do not yet realise this need to catch up!

Traditionally, the various phases from building concept to handover made use of multiple tasks, many of which were not directly coordinated. These included the drafting of drawings (usually numerous versions to cover various elements of the building and services); the preparation of multiple schedules of materials and assets; the writing of specifications for each aspect of the building (externally and internally); and production and construction detail documents. These would usually be supported by method statements for each stage of the building design and construction process.

Instead, BIM allows the detailed development of the building, virtually and in 3D, to provide a single holistic model. Any plan or specification can then be taken off the master model. A key benefit is that any modifications to the design will influence every affected element of the design.

The idea behind BIM can be traced back to 1975 with a concept referred to as a "Building Description System"[64]. A leading architect at the time, Charles Eastman, wrote a paper titled "The Use of Computers Instead of Drawings in Building Design". He discussed amalgamating different geometric objects for building design to create a single project, allowing the viewing of a given model from many different angles, combined with databases for the components of the building projects.

It is inevitable that the primary global supplier of computer-aided building design software – Autodesk- plays a lead role in BIM today. Their product is called "Revit". The benefits they cite on their website are:

- Improves efficiency and accuracy across the building project lifecycle, from concept to construction.

- Allows the automatic updating of floor plans, elevations, and sections.

- Performs routine and repetitive tasks automatically.

- Allows collaboration between the different parties (architects, engineers, project managers, etc.) to work together and deliver projects more efficiently and with fewer errors.

- Allows for specialist analysis of specific aspects of building design. Specifically, the system can permit, for example, structural analysis and performance of materials such as the types and arrangements of timber, concrete or steel solutions.

- Provide a complete auditing and information management tool, including automated schedules of components.

There are obvious benefits in BIM for the application of fire strategies. Strömgren[65] uses the Grenfell Tower Fire example in London (2017) and the subsequent review (the Hackitt Report) of the UK fire safety building regulations. The report acknowledged that BIM could improve the control and transparency of specifying fire safety and fire protection systems for buildings. In other words, a more robust audit trail, a factor identified as a failure in the Grenfell Tower Inquiry (still ongoing at the time of finalising this book). It is the central management of large amounts of information covering every asset and material used that is such a powerful tool.

Already, specialist software applications explicitly covering BIM-focused fire safety, engineering, and protection are being introduced. Some of these allow designers to allocate fire protection features to 3D CAD models. We can show the location of fire compartments together with their relevant ratings, the location of fire detectors with cable routes, and the layout and pipework location for sprinkler systems. A collaborative platform is provided using the web. Many similar applications will likely emerge.

One thing to note here is that BIM is becoming a global tool. A single standard building design methodology that can, at one instant, provide a uniform solution wherever the building is located.

Yet, architects and building designers will often need to revert to national requirements regarding fire safety. A concept of fire strategy formulation and verification that can neatly tie into the BIM phenomenon is, no doubt, desirable. This book aims to suggest a better process of providing fire strategies - on a global basis. Its goals are in tune with that of BIM.

7 Fire risk analysis and scenario determination: Introduction

Risk is a non-absolute concept.

So far, we have considered the background to fire safety and fire engineering. We have looked at the possibility of a global approach to fire safety standardisation. We have assessed the scope of fire engineering as incorporating many other areas of engineering and science. And we have considered new ideas that will influence the role of the fire engineer and the preparation of fire strategies.

To provide more relevant and holistic fire strategies, there is a need to evaluate the fire risk properly, both at a micro and macro level, by considering the building as a whole and within the community. The results can then be used to focus on the most crucial fire scenarios for further examination. The process can be the same for new construction projects as for a retrospective fire strategy evaluation of existing buildings.

Let me start with a personal perspective of fire risk analysis. Following the UK Regulatory Reform (Fire Safety) Order of 2005, many fire consultants carried out fire risk assessments to help building owners and occupiers meet their legal obligations. A relatively simple method was adopted by assessing various parts of a building and scoring a "risk" number by multiplying a probability factor with an impact, or consequence, factor. For example, if each factor were scored between

0 and 5, then the highest risk assessed number would naturally be 25. So far, so straightforward.

After undertaking a few assessments myself, I found that two scores, 9 and 12, were more common than any other risk score. This outcome revealed a natural inclination to score a 3 or 4 for both probability and impact. It was at this point that I questioned the subjectivity of the process. A room full of wastepaper combined with uncontrolled use of electrical equipment was a recipe for fire, and in such cases, the probability of fire would score 4 or 5. But how could you judge the consequences of a fire? This I found much more difficult to reason.

Indeed, an uncontrolled fire would mean the worst possible consequences where the building could burn down, and its occupants are put at high risk of death. So what would a score of 2 or 3 mean? Would it be that the assessor has identified that control measures are in place such that a fire would be limited to a single room or a specific zone of a building where the consequences are not that great? Or is it that the building is, for instance, old, falling apart, and unoccupied so that the consequences are relatively low?

The process is too simplistic and allows for a wide margin of error, being reliant on the views of the assessor on the day of inspection. I have even heard it said that an assessor in a good mood would score risks lower than if they were in a bad mood! Because of this, I prefer a more qualitative assessment where hazards and any control measures are described, with a possible risk rating given as high, medium, or low.

In my book "Fire strategies – strategic thinking", I point out that, in many ways, prediction of how a fire develops and its impact on a building is like predicting the weather. As with fire, the fundamental principles and patterns of growth and movement are known. What cannot be expected with accuracy are the results. Nevertheless, a range of risk analysis methodologies is used for the evaluation of risk.

The risk assessment process, as a subset of a risk management process, is described by ISO Standard 31000[66] and is illustrated in Figure 7.

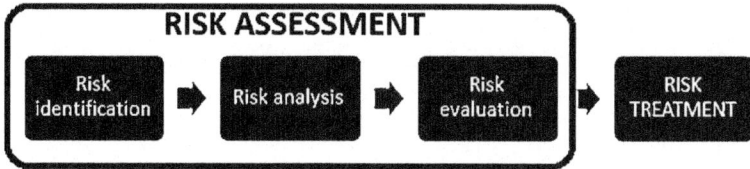

Figure 10: The ISO risk assessment process

In this Standard, risk management incorporates both risk assessment and risk treatment. Risk assessment is then broken down into three elements:

- Risk identification - the identification of those areas that offer a specific risk.

- Risk analysis – the analysis of how those areas identified will lead to certain unfavourable risk-based scenarios.

- Risk evaluation – a qualitative or quantitative evaluation of the level of risk found.

From the risk assessment process, appropriate levels of treatment can be applied. The Standard states that it is designed for use by people who create and protect value in organisations by managing risks, making decisions, setting and achieving objectives, and improving performance. It points out that organisations of all types and sizes face external and internal factors and influences that make it uncertain whether they will achieve their objectives. The Standard raises a few pointers:

a) It is iterative and assists organisations in setting strategy, achieving objectives, and making informed decisions.

b) It is part of governance and leadership and is fundamental to how the organisation is managed at all levels. It contributes to the improvement of management systems.

c) It is part of all activities associated with an organisation and includes interaction with stakeholders.

d) It considers the external and internal context of the organisation, including human behaviour and cultural factors.

Note that the ISO Standard is used to evaluate all risks and not those posed by a fire within a building. There are many guidance documents and codes that specifically cover fire risk. These will be described shortly. Let us begin by developing the elements given in Figure 7 for a fire safety context. The different stages in fire risk management can be described as:

Fire risk identification is the term given to a systematic process to understand how, when, and why fire could occur in each area of a building. This would typically be a qualitative process.

Fire risk analysis is the process of estimating magnitudes of consequence and probabilities of the adverse effects resulting from a fire in a building. The result of fire risk analysis is expressed either in qualitative, quantitative, or mixed terms depending on the type of risk, the purpose of risk analysis, how detailed the analysis is to be, and the information resources available. I earlier highlighted my preference for the qualitative method until a better quantitative way is found.

Fire risk evaluation follows from the analysis by judging the risk criteria to determine the appropriate or acceptable level of fire risk.

Fire risk treatment is assessing the efficacy of existing risk control measures and implementing additional controls where required.

Although fire risk assessment is considered just one aspect of the ISO risk management process, as explained, it has been used as the foundation of regulatory decision-making, i.e., the UK Regulatory Reform (Fire Safety) Order,

British Standard PAS 79[67] was purposefully produced to guide the risk assessment process and suggests nine steps:

1. Obtain information on the building, the building processes, and the people present or likely to be present in the building.

2. Identify the fire hazards and the means for their elimination or control.

3. Assess the likelihood of a fire.

4. Determine the fire protection measures in the building.

5. Obtain relevant information about fire safety management.

6. Assess the likely consequences to people in the event of a fire.

7. Assess the fire risk.

8. Formulate and document an action plan.

9. Define the date by which the fire risk assessment should be reviewed.

The above process is designed to be a practical guide for those undertaking fire risk assessments. There is a difference between a fire risk assessment for the benefit of legislation and one to assist with the development of a fire strategy, particularly for new constructions. In several cases, the former could better be described as a fire compliance assessment where the assessor compares their findings with the requirements of building regulations and standards and an assessment of the housekeeping and management arrangements *viz a viz* fire safety. The latter is likely to be more of a technical exercise where the risks could be determined and quantified by a team.

The use of a risk assessment process in identifying both the likelihood and impact of a fire can be categorised as one type of probabilistic risk assessment. Note that probabilistic risk assessment may be much more complex than a simple assessment based on a subjective judgement of two factors and then multiplying them together. It may use statistical techniques, which could be taken from data on historical fires. Assessments may use mathematical techniques such as regression analysis. Sensitivity or event tree analysis can be applied to provide a more robust conclusion by showing how a sequence can lead to specific outcomes.

Probabilistic risk assessments can be as straightforward or as complex as would be appropriate for a building, its occupancy, its process, and the objectives set for the fire strategy. PD 7974-7[68] is part of a suite of documents designed to support the performance-based approach described by British Standard BS 7974. It states that a probabilistic risk assessment can add value to traditional deterministic analysis and outlines acceptance criteria for the evaluation. It includes data to support a probabilistic risk assessment based on fire statistics, building characteristics, and reliability of fire protection systems.

Deterministic studies can help *determine* potential worst-case scenarios. To get to a point where this approach is practical, those undertaking the analysis would need to derive several significant scenarios and ensure that the chosen fire strategy can cope with them. Scenario evaluation is a topic that is covered later in this book.

A comparative study of risk is another approach that, as the term suggests, compares the risk of the building subject to the fire strategy with similar building and occupancy criteria. Comparative measures also include risk profiling.

Fire risk profiling can be a valuable and quick route to categorising a building and its occupancy. The profile can be based on a single aspect of a building or may cover several factors, such as:

a) Building size/complexity.

b) Building use.

c) Occupancy profile, including numbers, age range.

d) Mobility, sleeping/no sleeping, risks.

e) Average/worst-case potential for ignition.

f) Average/worst-case fire loading criteria.

g) Typical fire growth curve

Risk profiling tends to take a common-sense approach in that, for example, one hospital would have a similar risk profile to another hospital. One petrochemical plant would tend to be similar in risk level to another.

British Standard BS 9999, as covered in an earlier chapter, makes use of risk profiling and uses two measures to allow a specific building and its use to be categorised - expected fire growth rate and occupancy type. For example, an office building with occupants aware of the building and a medium growth rate will be classified with one profile. A sleeping risk, also with a medium fire growth rate, will have a higher risk profile.

The SFPE[69] has introduced a similar method of classifying risk by using a risk indexing system. Fire risk indexing systems are defined as an empirical model of fire safety and are particularly useful for the insurance application of risk-based upon hazard categorisation. The methods constitute various processes of analysing and scoring hazards and other attributes to produce a quick and easy estimate of relative fire risk. The process may start with a degree of subjectivity by using professional judgment and experience. Subsequent values can be allocated to selected variables, covering both positive and negative fire safety features. The derived numbers are then adjusted to arrive at a single value. In this way, we can compare assessments or against a chosen baseline standard. These principles are used later in this book to develop a fire strategy index system.

A risk index can be a single number measure of the risk associated with a facility[70]. Thus, for example, insurance rates are linked to fire risk indices.

Rosenblum et al.[71] provide a graphic view of the relative power and limitations of three general levels of risk quantification (Figure 8). Curves A, B, and C do not represent actual data points but are provided for relative illustration.

Curve A represents a probabilistic risk analysis using complex analysis of the hazards and possibly statistics. This analysis is described as the most accurate approach to defining risks, especially where the risk level is more sensitive. Significant resource investment is necessary to accomplish this task.

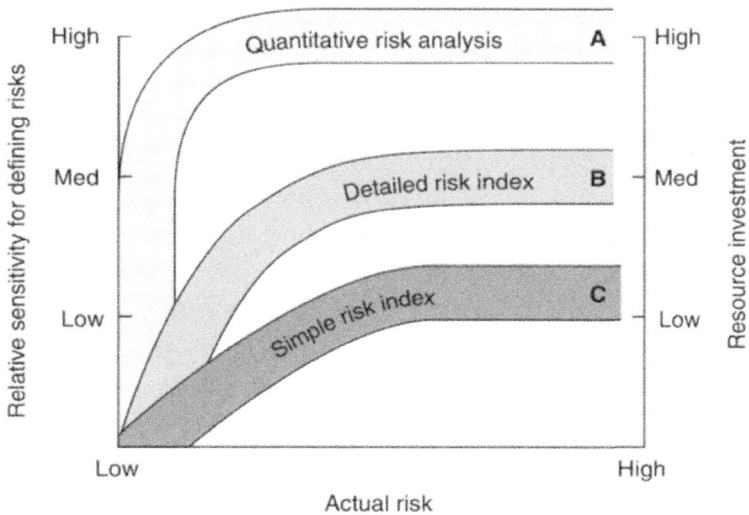

Figure 11: Risk indexing system and relative sensitivity (Rosenblum et al)

Curve C is a more simplistic method for fire risk identification. This method may be appropriate for high-risk worst-case loss situations but not so good for evaluating minor differences in risk level. This curve would best represent on-site fire risk assessments.

Curve B, as being central to the above curves, represents a method that may be used for normal circumstances.

A more complex and accurate assessment model will provide greater differentiation between lesser risks and improved overall accuracy. The trade-off for this approach is increased time and resources expended for the development, implementation, and data collection of a model or models.

An approach not dissimilar from the above was introduced by BS PAS 911. The approach also considers the resource commitment required to maintain a certain level of risk. Similarly, the same level of resources applied to two different risk profiles will leave different levels of residual risk. It is pointed out that risk vs.

resource commitment can be illustrated by curves not dissimilar to the supply and demand curves from economics theory. The curves are shown in Figure 9. The diagram illustrates that there is an ultimate risk/reward profile for all building types. We start from the idea that, with a minimal cost associated with applying fire safety and protection resources, there will be an understood residual level of risk. As resources are purchased and brought into the strategy, the level of risk will gradually be reduced until an optimum level of risk versus cost is reached. As more cost is invested into further risk mitigation provisions, the level of additional risk reduction will reduce until it reaches a stage where it may be judged to be acceptable.

Note that buildings with varying types and use will inherently have different risk profiles, so the risk/cost curve will be on a different plane—the riskier the profile, the more the curve moves to the right. The purpose of the curve analogy is to promote an alternative way to think about risk, particularly when the cost of the fire strategy is a measure.

It is worth mentioning the term ALARP which some risk assessment practitioners use for a practical assessment basis. "ALARP"[72] is short for *as low as reasonably practicable*. Another similar term, "SFAIRP", is the abbreviation for *so far as is reasonably practicable*. The two terms consider the practicality of risk management.

The judgement of when the term "reasonably practicable" can be utilised involves weighing a risk against the logistics, time and resource required to provide ultimate levels of risk control.

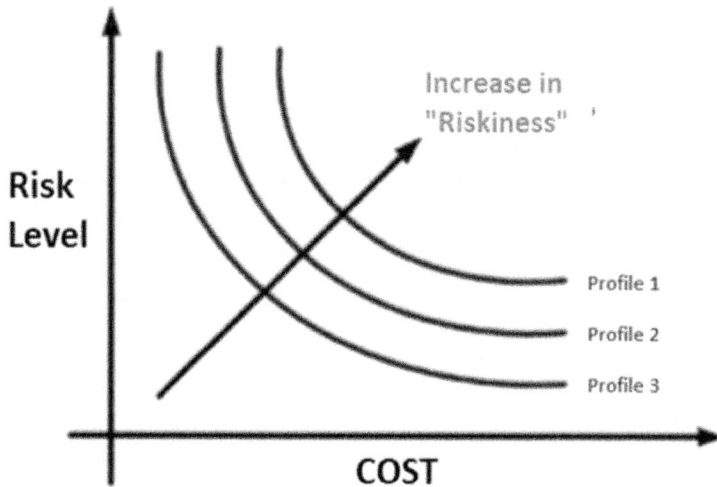

Figure 12: Risk/Resource profile curves (BS PAS 911)

Using the term "reasonably practicable" allows us to set realistic goals and provide practical solutions. This approach, therefore, accepts some flexibility over prescriptive requirements even though it probably also re-introduces some subjectivity. More formal decision-making techniques, including cost-benefit analysis, should be considered for high hazards, complex or novel situations.

The SFPE Handbook[73] defines fire risk as the product of the probability of fire occurrence and the consequence or extent of damage to be expected on the occurrence of fire. There are three identifiers:

a) Loss of or harm to something that is valued (e.g., life, property, business continuity, heritage, the environment, or some combination of these),

b) the scenario that may induce the loss or harm, and,

c) a judgment about the probability that the loss or harm will occur.

Fire risk is described as a weighted average of the risk values of each scenario, and it can be presented with the following formula:

$$FR = \sum_{i=1}^{n} P_{f_i} C_{f_i} \qquad \text{Eq.7.1}$$

Where FR is the fire risk, P_{fi} is the probability of occurrence of fire scenario i (per year), C_{fi} is consequences of scenario i; n represents the total number of scenarios.

The greater the number of possible scenarios that are chosen (n) the more accurate the determination of fire risk. Furthermore, the fire risk can be applied to several different objectives, such as the risk of occupancy deaths, the risk to the property itself, the risk to business, and the impact on the environment. Objectives' assessment is considered in a later chapter. Naturally, the determination of one or more scenarios for further assessment will be a crucial factor behind a performance-based approach when used with fire engineering.

ISO/TS 16733[74] introduces one approach in the identification and development of appropriate fire scenarios. Ten steps are to be followed from the location of a fire through to event tree analysis to final selection based upon probability, consequences, and the appropriate risk ranking (Table 3).

In a real sense, the number of potential fire scenarios is unlimited, particularly when we consider complex building environments. If the ten-step ISO approach is followed, then the more likely fire scenarios can be established.

Table 3: ISO recommended steps for determining fire scenarios

STEPS OF ISO/TS16733	COMMENTS
1.Location of fire	*Characterize the space in which fire begins as well as the specific location within the space.*
2.Type of fire	*Characterize the ignition, initial intensity, and growth of potential fires.*
3.Potential fire hazards	*Identify fire scenarios that could arise from fire hazards associated with the property's intended use or design.*
4.Systems impacting on fire	*Identify the fire safety systems and features that are likely to significantly impact the development of untenable conditions.*
5.Occupant response	*Identify actions that people take that can have a significant impact, favourable or otherwise, on the course of the fire or the movement of smoke.*
6.Event tree	*Construct an event tree that represents alternative event sequences from fire ignition to outcome associated with fire scenarios.*
7.Consideration of probability	*Estimate the probability of occurrence of each event using available data or engineering judgment.*
8.Consideration of consequence	*Estimate the consequence of each scenario using available loss data and engineering judgment.*
9.Risk rating	*Rank the scenarios in order of relative risk. The relative risk can be evaluated by multiplying steps 7 and 8.*
10.Final selection and documentation	*For each fire safety objective, select the highest-ranked fire scenario for quantitative analysis. Selected scenarios should represent the major portion of the cumulative risk (sum of the risk of all scenarios).*

The SFPE[75] utilises the ISO methodology and recognises that large numbers of scenarios can be prohibitive to most projects. A selection process using event trees, together with probability and consequence analysis, is the best way forward. It also highlights the use of a model developed by the National Research Council of Canada that uses a fire risk and cost assessment process. The model is called *FiRECAM*[76] and calculates the expected risk to life and the cost of a fire using hazard analysis of chosen fire scenarios.

The *FiRECAM* model reduces the number of potential fires into three basic types:

1. Smouldering fires where smoke only is generated.

2. Non-flashover flaming fires – generation of relatively small amounts of heat and smoke.

3. Flashover fires involving significant amounts of heat and smoke, with the potential for fire spread throughout the building.

Interestingly, the model utilises data from the USA, Canada, and Australia to consider the relative expectation of each type of fire. In this case, for apartment buildings, as illustrated in Table 4.

Perhaps not surprisingly, non-flashover fires are the most common, with a high degree of correlation over the three countries. The other fire types are also relatively consistent as a percentage. Possibly, given the statistics are from first-world countries, this uniformity could be attributed to those countries with similarly well-controlled fire safety laws. A more useful analysis would include a range of countries from all parts of the world. This simple evaluation points to the fact that a global methodology for fire strategy formulation would make sense.

Table 4: probability of fire types in apartment buildings[77]

FIRE TYPE	AUS (%)	USA (%)	CAN (%)
SMOULDERING FIRE.	24.5	18.7	19.1
NON-FLASHOVER FIRE	60.0	63.0	62.6
FLASHOVER FIRE	15.5	18.3	18.3

The New Zealand government has adopted an interesting approach to the subject of fire scenarios. In its standard ref C/VM2,[78] it recommends that, for a performance-based solution, ten separate fire scenarios should be considered for every fire engineering design. If any aspect of the engineered solution does not cater for these scenarios, then additional control measures will be required. The ten scenarios are:

1. Blocked Exit (BE): In this scenario, the document provides guidance on whether a single escape route is acceptable. This relates to the number of persons in a specific section of the building (room or floor). The figure of 50 persons is given as the maximum parameter where additional routes may be required. The figure is raised to 250 persons for sprinklered multi-storey buildings.

2. Unknown threat in an unoccupied room (UT): This scenario considers the threat to occupied areas from a fire that may start in an unoccupied area. Again, occupant loading of 50 is used as a benchmark. Other factors for consideration are automatic sprinkler systems, fire detection systems, and fire separation. One of the checks recommended is an ASET/RSET evaluation.

3. Concealed space (CS): Once again, the 50-person parameter is applied to determine whether concealed spaces, such as floor and ceiling voids, could create adverse conditions for evacuation. The dimensions and fire loading of such spaces are identified as defining criteria, together with active and passive fire protection.

4. Smouldering fire (SF): This scenario considers if a smouldering fire could adversely impact sleeping persons. The critical control measure here is the use of fire detection and alarm systems.

5. Horizontal spread of fire (HS): This scenario considers the spread of fire horizontally from one structure to the next. The premise here is whether the building is sprinklered or not. If not, it considers the dimensions of an "enclosing rectangle" boundary conditions of the building based upon fire loading per unit area and radiant flux.

6. Vertical spread of fire (VS): As with HS, this scenario considers how a fire could spread vertically via external walls. Of relevance here are whether there are sleeping occupants in upper levels, building height (10m criteria), and the use of façade materials.

7. Surface finishes and rapid-fire spread involving internal surface linings (IS): This scenario considers the impact of internal linings and points to relevant NZ standards for flammability.

8. Firefighting operations (FO): This considers access and firefighter provisions. Parameters for decision making include sprinkler protection, building boundary conditions, floor area, fire loading and expected radiation flux on arrival of the fire service.

9. Challenging fire (CF): This looks at potential worst-case scenarios in a normally occupied building. This scenario requires detailed consideration of ASET and RSET conditions to determine if the building has a suitable safety margin to allow evacuation before conditions become untenable. The "challenge" is whether the fire could create potentially challenging conditions for the building's fire safety systems.

10. Robustness check (RC): This scenario considers how a fire safety system's failure could impact life safety. Once again, this scenario is based upon the number of people affected (150 or 50 sleeping persons). The scenario should consider how any system could impact the ASET/RSET calculations and whether these will need to be revised.

Another interesting way of reviewing fire scenarios is to consider fire scenario clusters as presented by Jing Xin Chongfu Huang et al[79]. A 'fire scenario cluster' is a subset of fire scenarios that resemble each other. The universe of possible fires could be grouped into a manageable number of scenario subsets so that all the elements are present.

A fire scenario is a sequential set of fire events linked together by the success or failure of specific fire protection systems or actions. A fire event is an occurrence that is related to fire initiation and growth and may be impacted by existing control measures, occupant behaviour, and firefighter response.

In understanding fire risk analysis, a few fire scenario clusters can support the determination of frequency and consequence. The scenario cluster *types* could be a fire ignition scenario cluster, a fire automatic suppression scenario cluster and a fire behaviour cluster.

The clustering of similar fire scenarios is recognised by the NFPA in their guide NFPA 501[80]. This document divides fire risk assessments into four methodologies:

(1) Qualitative;

(2) Semiquantitative (likelihood) - where "likelihood" only is quantitatively assessed;

(3) Semiquantitative (consequences) - where "consequences" only is quantitatively assessed;

(4) Fully quantitative.

The document re-introduces Eq. 7.1 (in a slightly modified form) but also presents a derivative equation where multiple objectives may need to be considered, such as the inclusion of business risks.

This equation is shown below (Eq.7.2). The equation applies the summation of scenarios for one objective (j to m) with a secondary objective (i to n).

$$R_i = \sum_{j=1}^{m} \cdot \sum_{i=1}^{n} F_i C_{ij} \qquad \text{Eq.7.2}$$

Where R_i = Total risk, C_{ij} = Consequence of multiple losses, F_i = sequence frequency.

This derivative equation is an interesting idea and points to how the total risk can become convoluted and complex, especially when multiple objectives are considered. In the next couple of chapters, I will provide an alternative suggestion for both evaluating risks for specific fire scenarios and using the method to evaluate objectives and threats.

8 Fire risk analysis and scenario determination: An alternative approach?

A new idea need not reflect adversely on an old one.

Chapter 7 introduces several methodologies for evaluating fire risk and determining realistic and relevant fire scenarios. Some of the ideas follow from earlier concepts. Some challenge these concepts. Some introduce developments of earlier ideas.

Even with the wealth of information available, could a succinct methodology be provided that may assist in scenario determination in a globally consistent manner? In this chapter, I take some of the concepts introduced in the last chapter but modify their use. The intention here is to improve both objectives setting and threat analysis methods to provide a holistic fire strategy.

Let us start with the premise that the various probability/consequence formulas described in the last chapter are tried and trusted but may have certain flaws. Even with the techniques described, is it possible to fully understand the element of probability by treating it as a single variable? Let me suggest that, by separating the fire ignition probability from that of fire spread, a much more realistic probability factor can be created. I suggest this for the following reasons:

a) Fire ignition and spread require consideration of a different set of variables.

b) Fire ignition requires a more fundamental assessment of a small number of limiting conditions, namely local fuel and heat sources, in oxygen.

c) If fire prevention techniques are particularly good such that the probability of ignition is negligible or low, then further assessment of growth conditions is less important.

Assessment of the probability of fire growth will typically be more complex and require consideration of various factors, including potential routes for smoke and fire and existing limiting features and control measures. Following on, there are two probability factors:

Probability Factor 1: Fire ignition

We return to the fundamental elements of the fire triangle, namely, fuel, heat, and oxygen. If any one of these three parameters is non-existent, then, de-facto, we have no fire. In the real world, all three parameters exist to a lesser or greater extent. Accordingly, the probability of ignition is not likely to be absolute zero but may be thought of as negligible. Therefore, the scenario fire risk is simply an ignition (with a varying degree of probability) or non-ignition condition. The consequence, or impact, of a fire, is only relevant should the fire grow after ignition. In other words, the two factors are directly proportional. This idea can be represented by;

$$SFR_i \approx P_i \qquad \text{Eq.8.1}$$

Where SFR_i = Scenario Fire Risk (ignition) and P_i = Probability of fire ignition.

In my book "Fire strategies – strategic thinking" as well as British Standard Specification PAS 911, I describe that the risk or hazards associated within an enclosure will exist in one of two ways; the local environment of the enclosure (covering the static elements of the enclosure such as the structure, fabric, and fixtures/fittings) and the process (covering the dynamic environment such as people, equipment, operations, and so on).

A matrix (Figure 10) allowing the depiction of both the environmental hazards and process hazards is introduced to help understand this better. Every enclosure can be subjectively judged. The hazard matrix is primarily designed to assess how a set of hazardous conditions could lead to fire ignition and is not entirely relevant for fire growth when another set of conditions will be at play.

The quadrants are as follows:

Quadrant A: Low Environment Hazard / Low Process Hazard: An ideal situation presenting little risk to the building.

Quadrant B: Low Environment Hazard/High Process Hazard: There will be areas where processes are necessary, so a low environmental hazard is appropriate. Similar situations will be found in hotels, restaurants, and schools where the primary process is us and our foibles!

Quadrant C: High Environment Hazard/Low Process Hazard: A good example is in the case of buildings where the inclusion of drapes, bookcases full of books, and so on presents a hazard. However, the process within the area is controlled, possibly by good levels of fire safety management. Heritage buildings and museums are excellent examples of this type of risk.

Quadrant D: High Environment Hazard/High Process Hazard: In such situations, a fire is more likely to be an inevitability than a possibility. Waste management process plants come to mind in such cases.

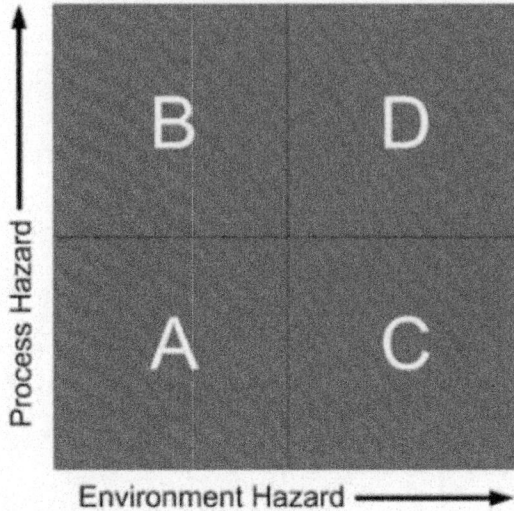

Figure 13: Hazard matrix (BS PAS 911)

By plotting each area, enclosure, or room on the matrix, a "hazard profile" can be made, rather like a hazard fingerprint for a building. In this way, fire engineers can focus on critical areas and devise a solution accordingly. Figure 11 gives an example of this in use. Thus, the probability of a fire ignition event can be considered from both the perspectives of the environment and processes.

Consequently, equation 8.1 could be further modified as follows:

$$SFR_i \approx \frac{(P_{ie}+P_{ip})}{2} \qquad \text{Eq.8.2}$$

Where SFR_i = Scenario Fire Risk (ignition), P_{ir} = Probability of fire ignition due to the environment, and P_{ip} = Probability of fire ignition due to the processes.

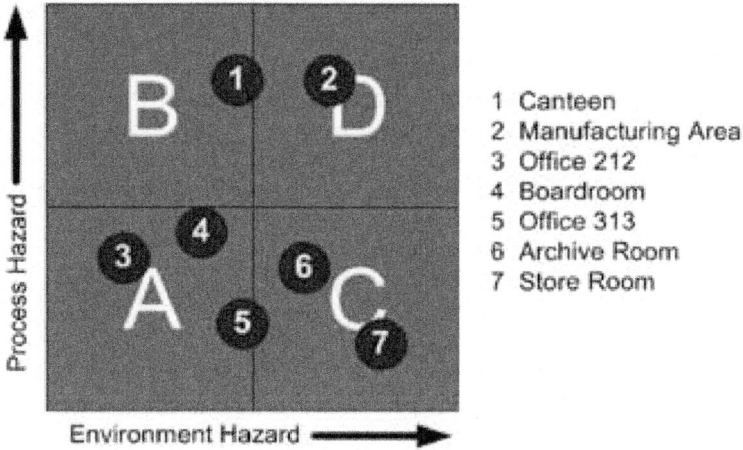

1 Canteen
2 Manufacturing Area
3 Office 212
4 Boardroom
5 Office 313
6 Archive Room
7 Store Room

Figure 14: Use of hazard matrix (BS PAS 911)

The Scenario Fire Risk (for ignition) is proportional to the mean sum of the two probabilities, given that, even if one of the hazard categories is near zero, the possibility of the other leading to a fire ignition is still there. The formula also supports the hazard matrix in that the highest probability will be where both the process and environment hazards are high.

Probability Factor 2: Fire growth

Following ignition, fire growth may be limited by both the local conditions and the forms of fire prevention and protection applied. Ideally, it is preferred that a fire is contained within the enclosure of origin or, better still, is eliminated before any threat can adversely impact people, property, the environment, etc. At this point, the consequences due to fire and smoke spread around a building should be analysed. A secondary formula is required to develop this idea as;

$$SFR_g = P_g C_g \qquad \text{Eq.8.3}$$

Where SFR_g = Scenario Fire Risk growth, P_g = Probability of fire growth, and C_g = The consequences of fire growth.

If both the ignition and growth phases are assessed collectively, the following combined formula can help determine the overall scenario fire risk. Note that, in this case. It will be the product of the two probability variables given that, without ignition, there will be no fire 0- and no fire spread:

$$SFR = P_g C_g \frac{(P_{ie}+P_{ip})}{2} \qquad \text{Eq.8.4}$$

For the ignition conditions, I have moved from proportional to absolute. This formula shows that we need to consider every scenario in terms of the probability that a fire can ignite and the separate probability of a fire growing and spreading. We can also establish the consequences of fire growth. If we consider several scenarios (n), then we can adapt this to the equation given by the SFPE as;

$$SFR_n = \sum_{i=1}^{n} P_g C_g \frac{(P_{ie}+P_{ip})}{2} \qquad \text{Eq.8.5}$$

Returning to Eq. 8.4, many quantitative risk assessments use a score from 0 to 5 for probability and consequences, leading to a maximum score of 25. Thus, for the SFR equation to be comparable, it will need to be divided further by 5. Therefore, the adjusted SFR equation will be:

$$SFR = P_g C_g \frac{(P_{ie}+P_{ip})}{10} \qquad \text{Eq.8.6}$$

Similarly, a range of risk scenarios using a scoring system of 0 to 5 can be represented as:

$$SFR_n = \sum_{i=1}^{n} P_g C_g \frac{(P_{ie}+P_{ip})}{10} \qquad \text{Eq.8.7}$$

Given that one of the objectives of Holistic Fire Strategies is to provide consistency in approach, why not provide a scoring guide? Table 5 suggests a scoring methodology for scores ranging from 0 to 5;

Table 5: Suggested scoring system for SFR

P_{ie}	No environmental ignition hazard within the enclosure = 0; Some environmental hazard (such as the use of low flammability linings, multiple electrical sockets, etc.) = 3; High environmental hazard (use of drapes, low-quality electrical systems, hanging fabrics, etc.) = 5
P_{ip}	No processes within the enclosure = 0; Some process hazard (such as computer installations, low-level fabrication, etc.) = 3; High process hazard (factories, processing, etc.) = 5
P_g	Fire contained at source = 1; Fire contained within the enclosure of origin = 2; Fire contained within fire zone = 3; Fire spread unrestricted within a section of the building = 4; Fire spread unrestricted throughout building = 5
C_g	Consequences of a fire v. low (minimal human presence, low-value assets, etc.) = 1; Consequences of a fire severe = 3 (some numbers of people present; high-value assets, some business interruption risk, some environmental risk, etc.) Consequences of a fire devastating (large numbers of people who may not know the building and may be sleeping or low mobility, significant asset risk, total business loss, and devastation to local environment) = 5.

I did contemplate the further sub-dividing of consequences into the four separate objectives of life, property, business, and the environment. However, this would overly complicate the concept at this stage.

In order to illustrate how the above can be used to formulate more considered scenario risk assessments, two example building profiles will be used - an underground metro station and a hospital. In both cases, I am applying some personal insight.

Metro Station

I will not use the example of a specific station but will apply my experiences at London Underground. A particular type of fire strategy is applied in all cases where the station is sub-surface. The following fire scenarios for evaluation are:

- Fire in the ticket hall area.

- Fire in an equipment room.

- Fire in passenger walkway.

- Fire at platform level.

The consequences of a fire are the same for all metro stations and could typically score C_g as a 4 or even 5. For larger complex stations, a score of 5 would be used, given the potential numbers of passengers affected and the limitations for escape, due to most people being sub-surface. The value would be relevant for all scenarios chosen. The probability scores are shown in Table 6. The *SFR* using the consequence factor of 5 is also shown in Table 6.

The calculus suggests that a fire in the ticket hall provides the highest risk scenario. The platform is also rated highly. In the latter case, the consideration of a train on fire scenario could further dramatically increase the risk. However, in this case, we are introducing another fire scenario into the mix. The probability of a train catching fire would first need to be separately assessed and then applied to the *platform on fire* scenario.

Table 6: Metro station: Suggested scoring for SFR

Location	P_{te}	P_{ip}	P_g	SFR	Commentary
Ticket hall	2	3	4	10	The Ticket Hall is an open area that interconnects with the remainder of the station without any physical barrier. The environment is mainly sterile, although the processes of ticket & vending machines and retail units are of higher risk. However, some units will be protected by fire suppression systems.
Equipment room	3	2	2	5	These rooms tend to be located within fire-resisting enclosures. The equipment can be regarded as fixtures, i.e., environmental risk but little in process risk.
Passenger walkway	1	2	4	6	These areas are primarily sterile, although a fire will be able to travel largely unrestricted.
Platform level	2	2	4	8	Platforms are sterile with no processes. Dust from train operations does create a slightly higher environmental risk. Some platforms may include fire barriers in the event of a fire, but it is assumed that they are not available in this case.

Hospital

The second example is of a hospital in the UK. Once again, no specifics, but let us assume that it is a multiple floor building with an underground car park. A feature of this hospital is a large single zone atrium which acts as the primary focus for visitors and staff to all building levels. The atrium encompasses a series of lifts, escalators, open stair cores, and glass and steel walkways.

The atrium includes interconnected lightwells. It is the atrium that allows regular access to all the wards. Let us also consider that an automatic sprinkler system protects all areas other than the central atrium.

The consequences of a fire C_g, should be regarded as suitably high in all occupied parts of the hospital (e.g., 4) and slightly lower for the car park (a rating of 3).

There are, of course, many potential fire scenarios in a working hospital. There are also numerous profiles within the building, from the highly controlled operating theatres and intensive care units to retail and restaurant units. Fire scenarios for special examination are:

- The central atrium.

- A typical hospital ward.

- The accident and emergency unit.

- The car park.

Table 7 shows the scoring methodology. Following the scoring, the A&E area potentially has the highest risk and is a target for specific evaluation. The central atrium is also potentially a scenario for further evaluation.

This chapter presents an alternative way to review risk by reassessing probability and dividing the component into the probability of both fire ignition and fire spread. Furthermore, the probability of ignition considers the independent variables of environment and process hazards within a given area.

The goal here is to provide a more focused consideration of risk factors, including a revised ISO-derived formula to determine a holistic "Scenario Fire Risk". This method will be applied to objectives setting and threat analysis as covered in the following chapters.

Table 7: UK Hospital: Suggested scoring system for SFR

Location	P_{ie}	P_{ip}	P_g	SFR	Commentary
The central atrium	4	4	2	6.4	As the atrium is large, let us assume it has been increasingly used for accommodating pop-up retail units, offices, storage, and even art galleries. These areas attract large numbers of visitors. This raises both the environmental and process ignition risk to a high level. Note that the atrium is sufficiently fire-separated from all other parts of the hospital by two sets of fire doors at multiple locations. Even though an atrium fire could conceivably spread throughout the space, the fire will be contained from all other areas. The height of the space and ventilation are unlikely to cause untenable evacuation conditions for evacuation.
Ward	3	3	2	4.8	A typical ward contains some environmental risk. Many hospital fire safety experts acknowledge that localised oxygen supplies increase the severity of an ignition event. The processes within an operational ward are also not negligible. Most wards are fire separated from other areas of a building.
A&E (Accident & Emergency)	3	3	3	7.2	A&E is generally treated as one evacuation zone. Such areas are typically located at ground floor level. In terms of risk, those found in a typical ward would be repeated here. The probability of fire growth is slightly greater due to the nature of the operations and the constant need to access other parts of the Hospital.
Car Park	1	4	2	3	The car park is in the basement. The environment is highly sterile, although the process (cars/car parking) could be regarded as high. The car park is fire separated from all other parts of the building.

9 A methodology for the incorporation of holistic objectives setting and threat analysis into a fire strategy: Part 1-Introduction

Limiting the scope of a concept is to limit its potential.

The purpose and scope of most fire strategies are to ensure that a building, or other forms of infrastructure, is compliant with national regulations, codes, and standards. This narrow approach can ignore specific risks and threats outside of those covered by such requirements. A broader consideration of factors not covered by legislation and the supporting regulations and codes may provide a better "holistic" application of fire safety and protection. The two areas of detailed assessment are:

Objective setting: Consideration of life safety, property protection, business protection, and the protection of the environment.

Threat analysis: Consideration of specific threats that may reveal issues not covered by the national base requirements.

The setting of objectives

It is easy for a fire engineer to be blinkered by the codes they use to achieve compliance. There is a need to encourage fire engineers to think about the broader issues at a sufficiently early stage in any construction project to ensure that the resulting fire strategy is truly holistic.

Acknowledging that most existing fire safety codes and regulations concentrate on the life safety of occupants, it is suggested that a holistic approach needs to consider much more.

A common perception of fire safety regulation is that it is all-encompassing, i.e., it provides satisfactory protection against fire for the building and its occupants. This is not the case. Fire safety requirements have habitually grown following a major fire that has involved multiple casualties. Lessons learned have led to revisions and updates to fundamental fire safety law, and this consequently follows through in supporting regulations and codes. The focus is on the saving of life.

Protection of physical assets, such as the building and contents, has typically been left to the fire insurance industry. Outside of any restrictions placed by insurers to obtain appropriate levels of insurance, there may be no requirement to consider anything other than adherence to the basic requirements of fire safety law.

There is an increasing realisation that many recent major fire events may have been avoidable had a more considered approach been taken for asset protection, business continuity, and protection of the environment. The following provides two relevant examples, the details of which can be found by a simple internet search.

The Liverpool Car Park Fire: A car park in an area of Liverpool, England, known as King's Dock, was the subject of a major fire that made headlines worldwide on the first day of 2018. The 1,600 capacity multi-level car park was subjected to a fire that entirely engulfed the building and destroyed all the cars on each of the levels. Estimates were that between 1,400 and 1,600 cars were lost. The car park is located directly next to a vast Arena – the Echo Arena, and by the Liverpool waterside. The Liverpool International Horse Show was being held at the Arena at this time.

This Arena has a total capacity of 11,000. Although everybody got out safely, some animals had to be rescued. It is thought that an old Land Rover was the cause of the fire when its engine burst into flames. The fire quickly spread from vehicle to vehicle until all the vehicles were consumed.

The actual cost of the fire in terms of property destroyed, loss of business for the arena and local area, and potential environmental damage to the nearby river was estimated to be around £60m, although a figure had not been publicly announced. Had broader fire safety objectives been adopted at the fire strategy formulation stages, most of this cost could likely have been avoided.

The Buncefield Fire: On 11th December 2005, at an oil storage terminal in Hertfordshire, England, a series of explosions impacted around twenty large storage tanks. A vapour cloud of evaporating leaking fuel had been ignited.

The fire was described as the biggest of its kind in Europe since the second world war. The black smoke cloud could be seen across much of Southeast England. This cloud contained hydrocarbons, which can be an irritant. Due to atmospheric conditions, the smoke cloud had also reached parts of the near Continent.

There were no fatalities, although 244 people required medical attention. As with the earlier example, it was the non-life safety impact that caused significant disruption and cost. This included:

- Hundreds of homes in the immediate area were evacuated, and about two thousand people had to be rehoused.

- The fire caused travel disruption by shutting main motorways until the fire eventually died down. Some local petrol stations reported long queues as people started panic buying.

- The oil terminal supplied 30% of Heathrow Airport's fuel, and because of the fire, the airport had to start rationing flights.

- The local industrial area was severely damaged, with six buildings designated for demolition and thirty more requiring major repairs.

- The local environmental agency had detected toxic substances affecting local water supplies due to the firefighting foam.

- The impact of fallen pollutants from the fire over parts of Europe had never been fully assessed.

- Then there are the direct asset and business disruption costs to the Terminal itself.

- Fines in the millions of GBP were attributed to some of the companies involved, but the actual final cost of the fire may be many multiples of the total fine values.

The above examples provide just two out of countless episodes where the fire strategy was not thoroughly thought through, and broader issues had been put to one side. Given the trend towards litigation alone, is it time that a fire strategy encompasses consideration of all relevant objectives?

The formulation phases of a fire strategy provide an ideal opportunity to consider much more than national legislation. The diagram below suggests that there could be up to 16 objectives for any and every fire strategy. There are four main objectives – Life, Property, Business, and the Environment. Each of these objectives is broken down into four sub-objectives to enable a focused analysis of performance objectives.

Figure 12 illustrates the fire safety objectives matrix, first published in the British Standard Specification PAS 911. The following descriptions are taken from my book "Fire strategies – strategic thinking" to assist in the understanding of why so many sub-objectives are included.

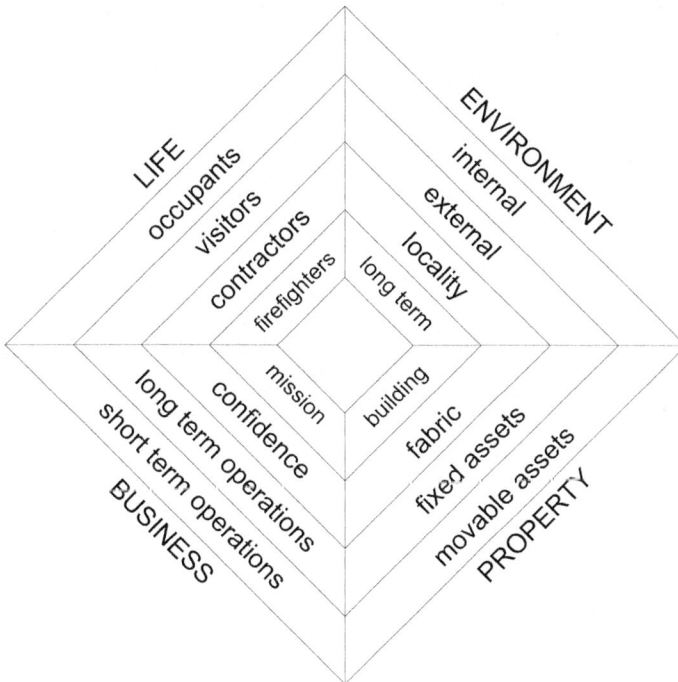

Figure 12: The fire safety objectives grid (BS PAS 911)

Life Safety

Let us first consider the key objectives of life safety and the building occupants specifically. Fire safety legislation is intended to ensure that the occupants of a building are, as far as possible, kept safe from a fire, and when required, can evacuate safely. Consequently, by meeting legislation, one could assume that the life safety objective has been covered, which in many cases may be correct. Even where prescription is used, the strategy should look deeper and consider the profile of occupants, their numbers, the level of disabilities, cultural or language issues, etc.

The second sub-objective is visitors. Visitors may have a different profile from occupants. It should be assumed that they have little knowledge of the building, and their requirements for being in the building may be varied. Visitors to a building may be a small proportion of the overall occupancy, such as in an office block. The public will also count as visitors, so a railway station, airport, museum, or sports stadium will have the majority population as visitors, with all their various profiles. Whereas standard occupancy numbers are generally known with some accuracy, the numbers of visitors may change at any time of the day. Visitors may also be unevenly spread out across the building; they may exist in high concentrations in specific parts of the building only. When determining means of escape, is this factor adequately considered?

The third sub-objective is contractors. Contractors are those persons who work on or in the building, whether for maintenance purposes or in construction projects. The main reason for inclusion as a separate sub-objective is that they will typically have different characteristics from occupants. They may or may not know the layout of the buildings. Their numbers may vary over a day or a week. The main reason for their inclusion has to do with evacuation times. When we usually consider evacuation, we calculate the travel time from the seemingly worst-case condition to a place of safety. This could be from the far end of a single means of escape, or an inner room, via another room, to an escape corridor. What is not usually considered is where the contractor may be when an alarm is raised. He or she may be in an attic area up a ladder, or in a confined space where movement is restricted, or even undertaking repairs to the roof. Their location will directly impact evacuation times and the methods used to raise alarms in certain areas.

The fourth life safety sub-objective is the firefighter. Very few countries now accept that firefighters know they are taking a risk and should be left to their own devices. Suppose the fire strategy is purely life safety of the occupants, and there is no requirement for firefighters to assist in any way. In that case, the safety of the firefighter can be discounted. For all other cases, firefighter safety should be paramount. Anything from specific provisions for accessing every building level to enhanced fire compartmentation over and above that necessary for evacuation will need to be considered.

Property Protection

The importance placed on the protection of property from a fire will vary greatly. Whereas the main objective of a life safety fire strategy is to ensure that all persons can be evacuated safely, the main objective of a property protection fire strategy is to limit the damage or loss of assets caused by a fire.

This requirement may include the summoning of professional firefighters to the scene as quickly and reliably as possible, or that automatic fire suppression systems are initiated. It could simply mean that the fire is contained by passive fire protection, or it may be a combination of tactics. Four sub-objectives are given. The first is the building itself. This requirement probably does not need explanation, and it is the building we usually think about when we discuss property protection.

The second sub-objective is the linings of the building. The most appropriate example here is that of heritage buildings. Sometimes the internal linings of the building are more valuable than the building itself. Wall and ceiling frescos are an excellent example of this. In these cases, it is not only flaming fire itself that the fire strategist needs to be aware of. Some frescos may be permanently damaged due to even small quantities of smoke.

The third sub-objective is fixed assets, i.e., those assets that may have intrinsic value but cannot be easily moved. By treating these assets differently from other property classes, a focused approach be taken. Examples here would include computer servers, manufacturing equipment, and test equipment.

The fourth sub-objective is movable assets and can include anything from computers to works of art. In such cases, the most valuable items may justify special consideration or dedicated arrangements.

Business Protection / Interruption

Despite worrying statistics about businesses going under after having a fire, many fire strategies rarely consider business protection or continuity. The way a business goes *kaput* is sometimes a gradual process that starts with the fire. Even if the employees are safe and the building is not destroyed, a business can still suffer. In the direct aftermath of the fire, the primary focus will be on the issues surrounding the fire. The focus may be removed from the business itself. Gradually, clients may start to look elsewhere, as the level of service they enjoyed is no longer apparent. Suppliers may find that dealing with the business has become more difficult. The fickle nature of business can often lead to a slow and inevitable decline.

The initial sub-objective is to consider how a fire will affect short-term operations today, tomorrow, and next week. Businesses may have established an effective business continuity plan that does not involve or require fire protection or fire safety management. Even so, it is worth asking the question, as some elements may not be instantly transferable and may warrant protection from fire and the effects of a fire.

The next sub-objective is long-term operations. Even if there is a quick fix to allow a business to survive in the short term, there may be an impact on operations in the longer term. It could be easy to switch manufacturing to another plant for a week, two weeks, or a month. What about next year? Will a fire eventually affect how the business works?

Confidence in the business because of a fire, or should we say, loss of confidence, can have a significant impact. It is this sub-objective that can often lead to the demise of the business. Those who operate daily in the public domain, such as a mass transit operator or the owners of a football stadium, will need to ensure confidence remains high in their ability to safeguard the public and control fires with minimal loss and disruption. Business confidence can be a shallow issue, but it may need to be taken seriously. In such cases, a fire strategy needs to lean towards high levels of fire prevention and fire safety management instead of relying purely on fire protection.

The final sub-objective is mission and is fundamental to the relevance of the business after a fire. If nothing else, consider how a fire can be so fundamentally damaging in so many ways that it may raise questions about the business itself. Imagine a fire safety teaching establishment suffering a major fire that destroys assets and highlights significant flaws in its fire strategy.

The Environment

The environmental consequences of building fire were rarely considered, but some national codes are beginning to acknowledge this issue. Let us begin with the internal impact of fire: the impact of a fire within the building. Whether the risk is from manufacturing processes, the storage of solvents or the chemical make-up of fixtures, a fire may release products that could prove a problem for the building environment. A proper understanding of this should help identify prevention or limitation techniques. Note that combustion products could contaminate anything within the building, such as manufacturing processes, stock, and consumables, making them unfit for use.

Then there could be issues about secondary contamination, such as the pollution of water supplies. What are the immediate and longer-term health and safety aspects relevant to using the building following a fire? What costs could be associated with any clean-up operations within the building?

Similar considerations should be given to the immediate zone around the building. The external impact of a fire may well affect all neighbouring buildings. It could be smoke damage or the impact of radiative heat on buildings immediately within the vicinity. It could be damage to cars in the car parks or to neighbouring processes that may not be sealed from the impact of a fire. With the increasing appetite for litigation in an increasingly competitive and unforgiving world, the potential is there for unforeseen fines from every direction. Let us also not forget possible health and safety implications to both persons and animals directly affected by the fire.

The next sub-objective moves to the broader impact on the locality. This includes the understanding of how the region in which the building is located may be affected by the fire. Considerations may need to look at how the fire plume, if not controlled, will affect the local community. Fires that reach tens of megawatts in size and lead to the release of airborne contaminants will be a problem. This could be compounded by the impact of various weather conditions and could lead to widespread issues. It may not even be the fire itself causing the problems but the impact of fighting a fire. The runoff from firefighting water could contaminate the local ground, rivers, and water supplies.

The final sub-objective is the longer-term impact of a fire. This issue may be hard to quantify and may be more subjective than the previous three sub-objectives. There will be a need to consider the more obvious scenarios and how they could pan out at a macro level, not just in months but in years. There is little doubt that some major fires have affected the local, regional, and national ecology. Examples that come to mind include the impact of a single fire incident on an offshore oil platform. The release of oil because of a fire can damage the local ecosystem for decades.

Threat analysis

Fire strategy documents sometimes explicitly state that "extreme events are not covered". What does this mean? Is this fire engineering code for *"we cannot guarantee that our fire strategy will protect you in more extreme fire events"*? In fact, any event other than a single fire event in any section of a building could be considered an extreme event.

As the world changes, the number and types of potential threats that could lead to a fire may similarly change. The opportunity to undertake a proper and thorough threat analysis could help identify possible fire scenarios that may otherwise have been ignored.

How often is there a formalised methodology applied to a proper assessment and analysis of how specific threats may impact the possibility of a fire event? A guide[81] produced by my old company Kingfell (titled KF912), is referred to below to provide some additional thought on this subject. The guide recognises that effective crisis management planning requires a proper assessment of threats that could lead to crises. It is also recognised that a crisis faced by many is that derived directly or indirectly by fire.

A threat assessment can help in the focus of what could happen given a specific set of circumstances. The guide allocates threats into six distinct groups. A threat assessment will help focus on the level of exposure an organisation may face to a crisis. An assessment can then be made of how each threat could manifest itself and how the outcome could impact the building or organisation. These six groups are described below.

1. **Intentional threats or actions:** The ability to understand how and where an intentional action or actions can lead to a crisis is paramount when managing that crisis. In some cases, even the threat of intentional action can lead to a crisis scenario without the action being instigated. It is in the planning for, and the reaction to, the threat where the severity and duration of the subsequent crisis can be minimised. Arson is a primary intentional threat.

2. **Accidents:** Accidents may range from minor mistakes to significant disasters. Some accident types may have a greater or lesser probability of occurring and a greater or lesser impact. Consequently, a full risk assessment can help identify the potential for accidents and how they

could impact individuals, the organisation, and the wider community.

3. **Environmental threats:** Environmental threats can usually be divided into two types; (1) Threats that are highly likely to re-occur, even if the frequency and impact cannot be readily ascertained, or (2) threats that may be totally or partially unexpected. Organisations based within "high risk" areas such as earthquake zones and flood plains are more likely to have taken steps to minimize the impact of such an event. There will be environmentally derived events that could directly affect or impact any building or organisation, some of which may not be readily envisaged.

4. **Economic threats:** Every organisation will be vulnerable to a greater or lesser degree by changes in economic conditions. There may be a direct impact on business trading or more subtle changes where the global economy or supply/demand relationships change over time. Arson could again emerge as a threat under this category.

5. **Threats to and from operations:** Operations, i.e., the day-to-day core activities of the building or organisation, can be affected by crises caused by internal and external situations. Operations could also introduce threats to the organisation.

6. **The actions of people:** Whether unintentional or not, the actions of people can lead to a variety of crises affecting both the organisation itself or outwardly causing threats to external persons or organisations. Many human resource agencies and experts will be able to provide information on this subject.

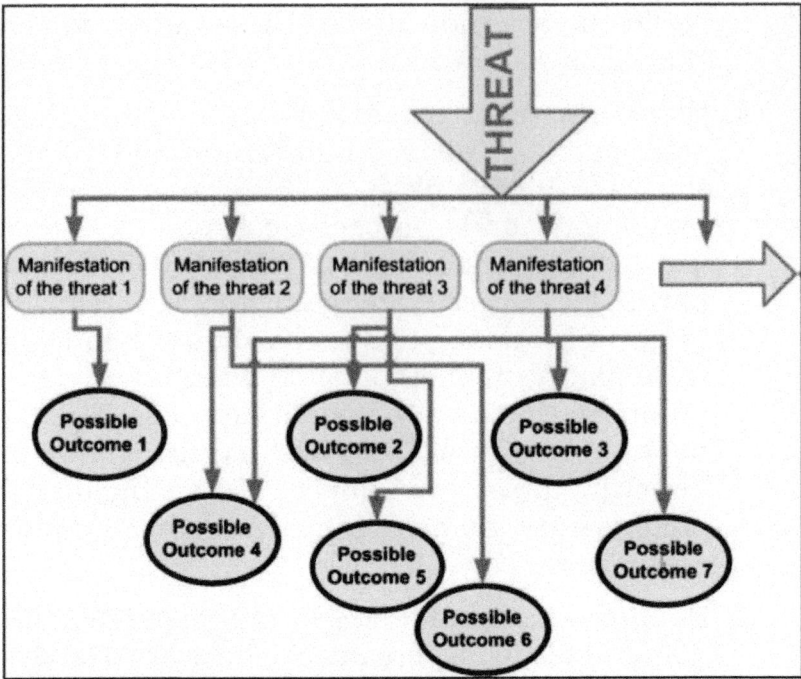

Figure 15: Threat/ manifestation diagram (Kingfell Guide KF912)

Figure 13 shows how threats may manifest themselves in several different ways. Each of these manifestations may then lead to one or more outcomes. Similarly, different manifestations of the threat may lead to the same outcome. Note that this flow chart is used for the wider aspects of crisis management. In this case, we are looking at the manifestation of the threat leading to one or more fire scenarios.

So, will threat analysis lead to better and more relevant fire strategies?

It could be said that a threat analysis will not change the need to meet with a minimum level of fire safety as determined both nationally and internationally. Nonetheless, it may focus on any issues that have not been adequately considered elsewhere.

As described above, threat analysis is based on a range of considerations. It is doubtful that generic codes can ever correctly consider the potential for fires and how they may manifest themselves. Because of this, and even with well-used fire safety standards, we still have notable fires where things do go wrong. Let us consider some examples. The first example is still evoking raw emotions in the UK, and an inquiry was underway when the following was written. Accordingly, I have tried to maintain a soft and objective evaluation but believe the example is worth analysis.

The Grenfell Tower Fire, London UK – 2017 [82]

On 14 June 2017, a fire broke out in the 24-storey Grenfell Tower block of residential apartments in North Kensington in central London just before one o'clock in the morning. It caused seventy-two deaths, and many more were injured. The building housed the more deprived of the local community within an otherwise wealthy neighbourhood.

Grenfell Tower underwent a major renovation during the years 2015 and 2016. The tower received new windows, a new heating system for individual flats, and new aluminium composite rainscreen cladding. The purpose of the cladding was to improve heating and energy efficiency and the external appearance of the building. It was this cladding that was believed to be the reason why the fire spread so quickly externally. It was suggested that many of the deaths were due to occupants remaining in their apartments as requested by the firefighters. This is a standard requirement for high-rise apartment blocks in the UK, known as the "stay put" policy or strategy. The policy relies on effective fire compartmentation of each residential unit.

At the time of writing this book, a government inquiry has yet to come to its conclusions. We do, however, have the suggestion that the fire started by overheating in the wiring within a fridge freezer in an apartment on a lower level of the block. Investigators had recovered the evidence from a small relay compressor compartment at the bottom rear of the fridge freezer. Note that the Fire Brigade had attended to this fire some hours earlier before the main fire.

If we review the incident against the six threat classes, could a preliminary analysis have prevented such an outcome? Of course, this is a retrospective evaluation but could help shine some light on how this method can be applied.

Intentional: A high-rise building housing a relatively poor community within a wealthy one could be exposing it to some degree of arson risk. The arson risk may come from other groups. A UK newspaper report in 2018[83] specifically highlighted that public officials feared that terrorists and arsonists could target high-rise homes that are still covered in combustible Grenfell-style cladding.

Accidental: Cooking appliance fires are described as the most significant cause of accidental fires in residential apartments[84]. This would not typically be a problem where the compartmentation between units is sound. Acknowledgement of this risk at a planning stage could highlight how the quality of fire separations is crucial to the fire strategy. We could, of course, stretch the accident factor to unintentional limitations in fire strategy scope, design failures for the external façade, and even aspects such as the lobby and staircase smoke control systems – all areas facing intense scrutiny by the Grenfell Tower Inquiry.

Environmental: If we look at this issue from a longer-term perspective, then the gradual climate warming could point to deficiencies in building design and increased susceptibility to the risk of fire. This fire occurred over a warm summer evening when open windows provided the only form of ventilation and the method by which an externally growing fire could quickly enter multiple units.

Economic: In this case, the findings regarding the use of unapproved and cheaper cladding materials have been a primary focus of the Inquiry. Economics may play a part in future fire strategies for residential buildings.

Operational: The internal operations of the tower block are probably less important than the operational issues in a fire event. This points to the safety management of those handling tower blocks, including evacuation and firefighting. Lessons could be learned from this incident that may impact the fire strategies of residential buildings going forward.

People: We should look at this from two respects. First are the actions of people that could cause or exacerbate the fire. Full compliance with the UK regulations should already allow for this. The second is the actions of people escaping. In this case, it is understood that they were advised to follow the stay-put policy, as is the strategy for high rise buildings. Unintentional as it was, this possibly led to many of the deaths.

Author's note: If any of the above has affected you, please let me know. Similarly, if you believe we need to learn from this tragic event, I would love to receive your comments. My contact details can be found at www.firecubed.com.

The Torch Building, Dubai, UAE – August 2017 [85]

A fire swept through one of the world's tallest residential towers in Dubai in August 2017. In a similar scenario as the Grenfell Tower fire, the fire took hold of the external parts of the seventy-two-storey building and grew rapidly. In this case, there were no fatalities. All the occupants (nearly 500 persons) could evacuate using the internal means of escape as intended. The

Dubai Civil Defence concluded that the cause of the fire was that a discarded cigarette was thrown into some plants on one of the balconies, which then ignited.

The UAE revised its building safety code in 2013 to require that cladding on all new buildings over 15m is fire-resisting, but older buildings were exempt from this requirement. UAE media had reported that most of Dubai's high-rise buildings use cladding panels with thermoplastic cores. Panels can consist of plastic or polyurethane fillings sandwiched between aluminium sheets.

As with Grenfell Tower, if we review the fire against the six threat classes, could a preliminary analysis have prevented such a fire? A similar fire scenario to Grenfell Tower but with a different outcome.

Intentional: This high-rise building is in an affluent area with high levels of security. Arson-type fires are less likely. Although the Middle East Region has seen more than its share of terrorist attacks, the UAE has fortunately not suffered in this respect.

Accidental: A similar accident profile as Grenfell could be attributed here. The detection and fire compartmentation appeared to meet the strategic objectives. Nevertheless, the balcony was exposed as a weak point, and a cigarette led to the fire, which exposed the properties of the external cladding.

Environmental: The tower block was relatively well sealed against the external environment. There is full air-conditioning. Even with people using the balcony, they are likely to keep the door to the balcony closed to maintain comfortable conditions inside.

Economic: It may be perceived that a wealthy nation would expect that the best materials were used in construction. As with all economies, there will be those who will seek to maximise profit.

Operational: The management and occupants of Dubai buildings tend to adopt a solid adherence to operational procedures in a fire. Dubai Civil Defence is also in a powerful position to attend and assist.

People: As identified above, occupants are highly likely to adhere to building management requirements. The success of the evacuation bears this out.

Major Warehouse Fire – Springfield, Ohio 2006 [86]

What started as a traffic accident turned into a massive fire in Springfield. A storage building was destroyed by fire after a van crashed into two telephone poles, sending the wires onto the top of the building's metal roof and leading to an unfortunate chain of events. The storage building and its contents were destroyed. Firefighters had to wait more than half an hour before putting water on the blaze for safety reasons. They could not fight the fire until the power to the electric wires was shut off.

A review provides the following conclusions:

Intentional: Many warehouses are susceptible to an arson attack. In this case, the area is primarily agricultural, and arson is judged to be relatively rare.

Accidental: It could be argued that this is a freak accident leading to a fire and could not be readily envisaged by a prior threat analysis. The lessons learned could be helpful for other threat analyses.

Environmental: Although not deemed initially relevant, a proper study may reveal issues such as tornadoes, which are becoming more commonplace. Could not a tornado lead to a similar occurrence?

Economic: Although there were seemingly no issues from an economic perspective. The absence of a sprinkler system (or at least an operational sprinkler system), for example, could be due to economic constraints.

Operational: Warehouses have a range of issues that could lead to a fire. The operations and storage requirements would probably have been assessed for fire risk by insurers or other stakeholders.

People: The fire was caused by the indirect actions of the driver of the van. This quite rightly could not be predicted for a fire strategy. It is the actions of people after the accident that could reveal issues that may have greater relevance.

Using a threat analysis model retrospectively may not reveal the true nature of how the model can be used, however the commentary for each of the above events does provide some thought, especially when broken down into the six distinct areas.

In summary, this chapter provides a background to the strategic assessment of objectives and of threats as they could apply to the formulation of fire strategies. The following chapters describes a proposed methodology to consider such issues in a structured manner and to improve the scope and coverage of our holistic fire strategy.

10 A methodology for the incorporation of holistic objectives setting and threat analysis into a fire strategy: Part 2-Identification

We can only solve a problem when we know what the problem is – and if there is a problem.

This chapter provides a framework for the evaluation of both objectives setting and threat analysis. The process is shown in Figure 14. There are three separate sub-processes.

Process A follows a sequence of tasks designed to evaluate objectives. From the evaluation, the critical objectives and associated scenarios can be identified. If it is believed that all scenarios are covered by the scope of the fire strategy when formulated following national requirements, then no more action is required. If not, then those scenarios will be assessed by process C.

Process B follows a sequence similar to Process A but covers threat analysis.

Process C analyses the chosen scenarios. It categorises them into possible clusters where appropriate. A scenario fire risk assessment is undertaken, and key scenarios that require special consideration are identified, with the purpose of applying additional control measures into the fire strategy.

Processes A and B are described in this chapter. Process C is covered in the next chapter.

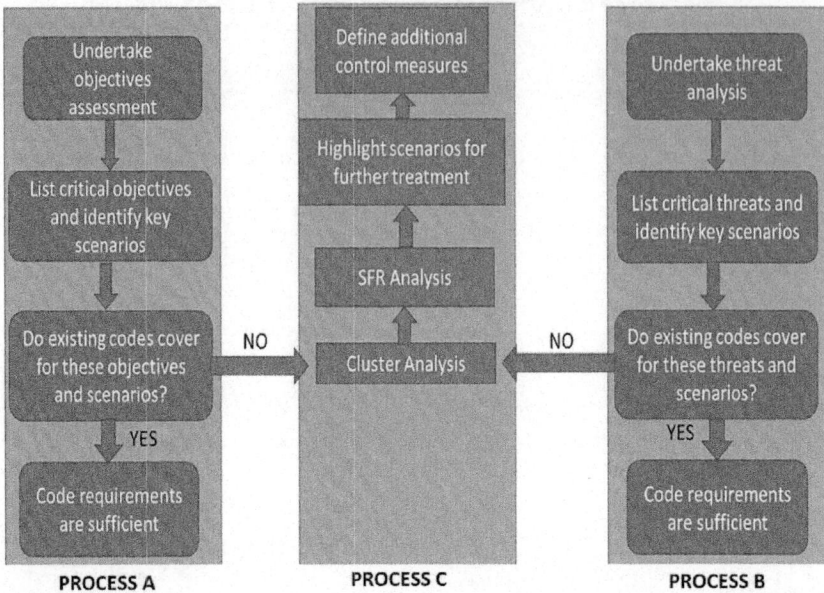

Figure 16: Process chart for the evaluation of objectives setting and threat analysis

Methodology for improved objectives setting and evaluation

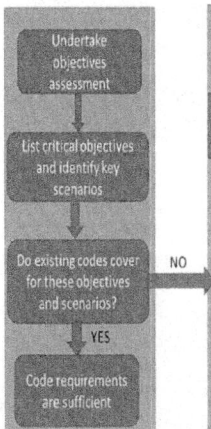

The importance and relevance of objectives assessment for a fire strategy is proposed by following a qualitative assessment process. As highlighted in the previous chapter, the primary objectives (Life Safety, Property Protection, Business Protection, and Environmental Protection) can be evaluated by assessing each sub-objective in turn. From this, three steps can be followed to derive a simple summary of relevant control measures.

STEP 1: Identification of the relevance of objectives, and sub-objectives, by categorising them accordingly. These are:

Category A: Objectives deemed highly relevant and should be a critical feature of the fire strategy.

Category B: Objectives deemed to be relevant but are not critical.

Category C: Objectives that are not appropriate or are irrelevant.

This exercise could best be undertaken at a meeting with all relevant stakeholders. This exercise would focus on issues that may just not have been thought about if adherence to national legislation and codes was the *only* objective. This will particularly be the case where an objective is intentionally entered as a Category C. It would also be worthwhile for the reasoning to be documented. The assessment could make use of a table, as shown in Table 8.

Once Table 8 has been completed in full, we must now focus our attention on those objectives categorised as either A or B. For objectives categorised as A, these are features of the strategy deemed sufficiently necessary to be catered for in the final fire strategy. These may be covered by national fire codes or may be outside the scope.

For category B responses, it is entirely likely that these are viewed as *nice to have's* but are not regarded as critical. There are two alternative methodologies to the treatment of these responses. The first is a secondary review, where they are re-allocated as category A or C responses. The second is to determine if scenarios that could impact the B category are sufficiently covered by either those in Class A or by the current codes being used.

Table 8: Objective setting of categories

Sub-objective	Category	Reasoning
L1: LIFE SAFETY: Protection of Occupants		
L2: LIFE SAFETY: Protection of Visitors		
L3: LIFE SAFETY: Protection of Contractors		
L4: LIFE SAFETY: Protection of Firefighters		
P1: PROPERTY PROTECTION: Building		
P2: PROPERTY PROTECTION: Fabric		
P3: PROPERTY PROTECTION: Fixed Assets		
P4: PROPERTY PROTECTION: Movable Assets		
B1: BUSINESS PROTECTION: Short Term		
B2: BUSINESS PROTECTION: Long Term		
B3: BUSINESS PROTECTION: Confidence		
B4: BUSINESS PROTECTION: Mission		
E1: ENVIRONMENTAL PROTECTION: Internal Impact		
E2: ENVIRONMENTAL PROTECTION: External Impact		
E3: ENVIRONMENTAL PROTECTION: Impact on Community		
E4: ENVIRONMENTAL PROTECTION: Longer-term Impact		

STEP 2: Once the list of critical objectives is complete, an appreciation of the most likely fire scenarios that would impact the critical objectives should be undertaken. An appropriate three-part process is shown in Figure 15.

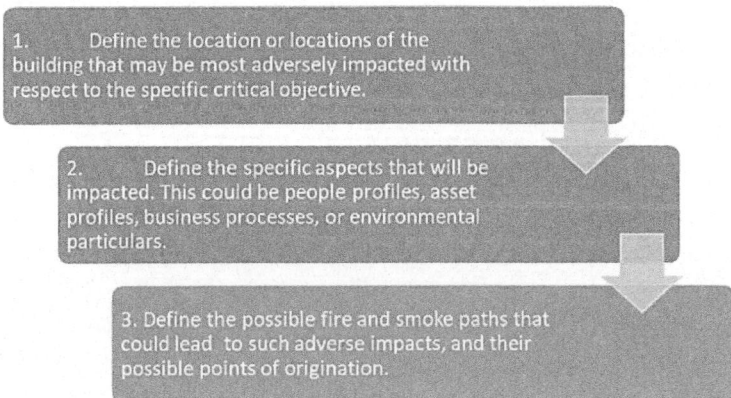

1. Define the location or locations of the building that may be most adversely impacted with respect to the specific critical objective.

2. Define the specific aspects that will be impacted. This could be people profiles, asset profiles, business processes, or environmental particulars.

3. Define the possible fire and smoke paths that could lead to such adverse impacts, and their possible points of origination.

Figure 17: Three-part critical objective process

STEP 3: At this point, we may have a list of scenarios for evaluation. We must establish whether these have been incorporated into the existing fire strategy and covered by the relevant codes and regulations. Issues may be, for example, critical fire and smoke paths may not be covered by the existing control measures. Another could be that existing smoke control measures may not cater for the identified scenarios. This assessment need not undertake a risk analysis as that will be covered later. The exercise should be undertaken by those appropriately knowledgeable of national requirements.

If all scenarios are thought to be covered by the proposed strategy, no more analysis is required. If not, then the analysis moves to process C.

Worked Examples

Below are a couple of worked examples. One assesses a metro station as used in the previous chapter. The second example is based upon a large school building.

Metro railway station

Following on from an example used in the previous chapter, this fictitious underground metro train station may identify the following objectives as critical (Step 1):

L1, L2, L3, L4, P1, B1, B3, E1.

The metro station fire strategy is supported by unique internal standards over and above the basic requirements of legislation (Step 2). It is therefore believed that all L category objectives will be covered together with P1. It is agreed that B1, B3, and E1 require additional consideration. The process makes use of the three-part analysis to derive the following scenarios.

Category B1: The metro operator wishes to avoid short term operational closure. This could affect the station and close the whole metro system until a fire has been safely extinguished and any damaged carriages have been removed from the permanent way (track). The critical location for consideration is at the platform level. Both persons on the train and the platform will be affected. Attention also needs to be given to the trains coming into the station and the equipment located at platform level. The platforms themselves are sterile, so a train on fire scenario or equipment rooms at the platform level will be deemed relevant.

Category B3: An effective metro system is only viable if passengers have confidence in the operations. An uncontrolled fire situation that occurs on a semi-regular basis will undermine this confidence in any part of the network. Therefore, where at all possible, small fires in any part of a station should be tackled quickly and efficiently.

Category E1: It is acknowledged that, from experience, fighting a fire has led to firefighting water seeping into local subterranean river networks feeding into primary river routes and contaminating these rivers. This was worse at the lowest levels of the station and, naturally, increases substantially as the fire to be tackled increases in size. Therefore, the scenario for consideration is a fire starting at the lowest level and allowed to grow to a size where manual firefighting using water supplies is deemed necessary.

From the above analysis, the following chosen scenarios are included for special consideration and were not covered by the current scope of the strategy:

Scenario 1: Train on fire arriving at the station with the objective of minimising downtime of operations.

Scenario 2: Equipment rooms at platform level experiencing a fire with the objective of minimising downtime of operations.

Scenario 3: Uncontrolled small fires at any part of the station with the objective of preventing such fires or minimising their impact.

Scenario 4: Uncontrolled fire at low levels (platforms) requiring manual firefighting may lead to local environmental contamination of areas, including subterranean rivers.

School building

This school, built over five floors, was constructed in the 1950s and serves the local community. A second school is located nearby. No formalised fire strategy has ever been prepared for the building. However, following a meeting with the local authority and other stakeholders, the following objectives were deemed as critical:

L1, L2, L3, L4, P3, E4.

It was identified that L1 and L2 were covered by the local building regulations and the corresponding fire safety standards at the time of the original building construction and are still relevant. The following objectives are retained for examination using the three-part process given above.

Category L3: It has been identified that the school has a sub-basement and roof that could be difficult to escape from in an emergency.

Category L4: The school does not have any special provisions to separate firefighters from risk areas on higher floors in a fire. The main staircase leads directly into open plan risk areas on the 4th and 5th floors.

Category P3: Although the school was fitted with an EU funded multi-million-euro computer suite, there were no systems to detect or suppress fires.

Category E4: The ceiling tiles throughout the school are manufactured from asbestos-based materials.

Experts have informed them that a major fire could result in the release of asbestos fibres over a large area, resulting in contamination, clean up, and long-term health issues within the vicinity.

From this analysis, the following fire scenarios for further examination were identified:

Scenario 1: A fire that could create untenable conditions for evacuation from sub-basement and roof areas.

Scenario 2: A fire on the 4th and 5th floor requiring firefighter assistance will currently expose them to the direct impact of a fire.

Scenario 3: A fire affecting the new computer centre that could lead to asset loss.

Scenario 4: A fire of sufficient severity to degrade the asbestos ceiling tiles and causing the release of asbestos fibres.

These scenarios will be further examined in Process C.

Methodology for threat assessment

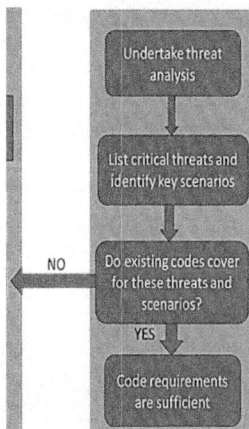

If we believe that a proper threat analysis at the commencement of fire strategy formulation is valuable, there should be an appropriate methodology. It is suggested that the following qualitative approach is appropriate to help identify if additional measures are required.

As described in the last chapter, threats are divided into six distinct groups. We first need to determine the threats that could potentially or conceivably lead to a fire event (Step

1). This determination can be made following responses to a series of questions about the building and the occupant profiles of those who will be, using that building. All relevant stakeholders should consider these questions. Once we have a condensed list, we will need to evaluate the potential fire scenario or scenarios from such an event (Step 2).

Our final step is to determine if the base legislation and codes already cover the fire scenarios or whether additional control measures would be recommended. The remaining scenarios would then be assessed by process C. Note that the control measures may possibly lay outside of fire safety and protection in the case of threat mitigation.

STEP 1: Identification of threats

Intentional threats or actions: The most apparent fire-related outcome of intentional action is arson. Nonetheless, an intentional action that is not arson may also lead to a fire event. Examples of this are the malicious treatment of equipment or materials. The following questions may prompt a realistic assessment of the likelihood of an intentional fire event. If the answer to any of the following is affirmative, then the risk should be considered conceivable.

a) Is the building located in an area that is known for arson attacks?

b) Do the primary or secondary uses of the building increase the possibility of arson type events?

c) Does the organisation or individuals located within the building support any political or religious causes that could be deemed to conflict with outside individuals, groups, or organisations?

d) Could any of the occupants be associated with national or sector-based activities that could offend other individuals, groups, organisations, or nations?

e) Could the actions of occupants, or a neighbouring organisation, or an organisation based in the same sector, lead to an increased risk of intentional action?

Accidents: Accidents may range from minor mistakes to disasters. A full risk assessment can help identify the potential for accidents and how they could lead to a fire event. The following considerations could help in identifying significant threats:

a) Has there been an assessment of possible and potential serious accident types that could be attributed to the building, its fixtures and fittings, its uses, or its occupancy profiles? Could any of these accident types lead to a fire event?

b) Is there any relevant data on past fire events for the building and occupancy profile that have led to a fire event? This could include industry sector data on the type and frequency of accidents.

c) Has the organisation made any changes in operations or structure that could lead to accidents that may not have previously occurred and potentially result in a fire event?

d) Is there any potential from accidents caused by external personnel or organisations that could lead to a fire event?

e) Are there any "chain reaction" accident types in the organisation where one accident could lead to further incidents resulting in one or more simultaneous fire events?

Environmental threats: It is increasingly likely that environmental threats could directly or indirectly lead to a fire outcome, amongst many other outcomes. Environmental threats can usually be divided into two types:

- Threats are highly likely to re-occur, even if the frequency and impact cannot be readily ascertained.

- Threats that may be totally or partially unexpected.

The following considerations could help in identifying significant threats. Note that the list is not exhaustive.

a) What are the known environmental threats that could impact the building or its surrounding infrastructure? How have these been accounted for?

b) What are the potential environmental threats that could impact the organisation yet have not been accounted for?

c) How could longer-term climate changes affect the type, frequency, and impact of the known or potential environmental threats that could lead to a fire?

d) Are there any activities undertaken within the building that could increase the risk of an environmental threat that, in turn, could lead to a fire event?

Economic Threats: Economic threats may directly impact business trading, or there may be more subtle changes where the global economy or supply/demand relationships change over time. Typically, it would not usually be envisaged that an economic threat can be directly linked to the cause of a fire event. However, there may be indirect causes, such as the reduction of fire safety budgets. This could lead to the lowering of both fire safety management efficacy and possibly leave critical fire protection systems at a level that they may no longer be fit for purpose. As a result, the risk of fire and fire not being controlled by the installed control measures will be increased. Consequently, there is only one question, and that is:

a) Could the budget to maintain the fire strategy be impacted by changes in the economic fortunes of the building owners or organisation(s) occupying the building?

Threats to and from operations: Operations, i.e., the day-to-day core activities of the organisation, can be affected by incidents caused by internal and external situations. Operations could also introduce threats to the organisation. In this case, we are primarily concerned about operations that could lead to a fire event. In this case, considerations should include:

a) What are the critical elements of failure in operations that could lead to a fire event? Have they been risk assessed? How have these risks been mitigated to date?

b) How will a partial or complete power failure impact the organisation in terms of increasing the risk of fire and reducing the availability of fire protection systems?

c) How resilient are the existing IT and communication systems related to supporting the fire strategy? What would be the impact of partial or total failure?

The actions of people: Whether unintentional or not, the actions of people can lead to a variety of fire-based events affecting both the organisation itself or outwardly caused threats to external persons or organisations. The following considerations could help in identifying significant threats:

a) Does the corporate culture of occupants of the building allow or even encourage attitudes that could lead to actions taken against the organisation's best interests? In terms of a fire event, this could be in the form of arson.

b) Is any organisation occupying the building beholden on one or more individuals who ensure that the fire strategy is maintained? Would the absence of such persons' impact any aspect of the fire strategy?

c) Could the organisation be associated with national or sector-based activities that could cause offence to individuals, groups, organisations, or nations and could, inadvertently, lead to an increased risk of fire?

STEP 2: A thorough analysis of threats may, or may not, result in an identification of fire events. The scenarios may differ for each of the threat classes, or there may be common elements where different threat classes may lead to the same or similar scenarios. Scenarios can be recorded in a tabular format by identifying threat classes and relevant scenarios.

STEP 3: As with the objective's evaluation process, once the remaining threats and related fire scenarios have been identified, these can then be evaluated in terms of the original scope of the fire strategy. The exercise should be undertaken by those appropriately knowledgeable of national requirements.

If all aspects are believed to be covered by the proposed strategy, no more analysis is required. If not, then the analysis moves to process C.

Worked Examples

For comparison purposes, the same two examples used in the objectives' analysis will be used here.

Metro railway station

Intentional threats or actions: The city in which the metro system operates has experienced heightened levels of arson. Consequently, the impact of an arson attack in a station is deemed accurate. The most likely fire scenarios will be in the passenger walkways, on platforms and in the train carriages. Upper floors are usually well policed.

Accidents: The Operator already has an extensive accident control system which includes accidents leading to a fire.

Environmental threats: As identified with objectives setting, fighting a fire has led to firefighting water seeping into local subterranean river networks and contaminating these rivers. The scenario has been suitably illustrated via the objectives' assessment.

Economic threats: The system is government-funded such that any economic issues affecting the fire strategy are deemed irrelevant for this assessment.

Threats to and from operations: An operational control system identifies every process and action arising.

The actions of people: One issue that has been identified is passengers becoming confused in an emergency and unaware of what to do. This is deemed mainly an issue at sub-surface levels and can lead to delays in evacuation.

From the above analysis, the following chosen scenarios are included for special consideration:

Scenario 1: Arson on a train leading to a "train on fire" in a running tunnel or entering a station.

Scenario 2: Arson in passenger walkways and at platform level.

Scenario 3: Fire in sub-surface passenger routes and platform leading to passenger confusion and delayed evacuation times.

School building

Intentional threats or actions: Arson is a cause for concern at many schools. Given that the school is locked overnight, the main risk is from fires being lit externally, such as waste bins.

Accidents: The chemistry lab has experienced small fires in the past caused by accidents during school hours. The fires have always been controlled using portable fire extinguishers.

Environmental threats: There are no environmental issues that can lead to a higher risk of fire.

Economic threats: School budgets have been reduced year on year. This has impacted the ability to maintain fire protection systems, notably the manual fire alarm system

Threats to and from operations: The chemistry lab has been identified as a potentially high fire risk area.

The actions of people: Arson has been deemed the most likely outcome of a grievance from a pupil or parent.

From the above analysis, the following chosen scenarios are included for special consideration:

Scenario 1: Arson by lighting waste bins located next to the school building.

Scenario 2: Failure of existing manual fire alarm system due to lack of maintenance.

Scenario 3: Uncontrolled fire in chemistry lab.

The next chapter will describe how the scenarios will be treated to ensure that both critical objectives and threats are appropriately considered for a holistic fire strategy.

11 A methodology for the incorporation of holistic objectives setting and threat analysis into a fire strategy: Part 3-Analysis

The most effective methodologies take a complex set of data and lead to a simple and effective conclusion.

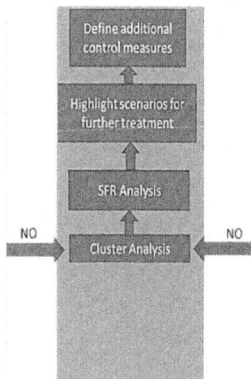

As described in the previous chapter, by undertaking a logical assessment of both objectives and threats, a list of specific fire scenarios may be found that fall outside of the considerations covered by national legislation, regulations, and codes. This chapter provides a framework for evaluating the chosen fire scenarios, prioritising them based upon risk, and determining suitable control measures. The process involves the following steps:

a) Assess the fire scenarios against a form of cluster analysis. This will help in identifying common features of the scenarios.

b) Undertake a scenario fire risk analysis of each of the remaining scenarios.

c) Determine suitable additional control measures over and above those required by the national requirements.

Cluster Analysis

Previously, the idea of using a cluster evaluation technique for fire scenario assessment was introduced. It suggests that a 'fire scenario cluster' is a subset of fire scenarios that resemble each other. This will help reduce scenarios into a manageable number. In the process of understanding fire risk analysis, three fire scenario clusters are suggested as essential to support calculations of probability and consequence:

Ignition Cluster: This covers scenarios where the ignition point of a fire may be at the same or similar location.

Growth Cluster: This covers the means by which fire and smoke can spread throughout part or all the building.

Impact (Consequence) Cluster: This covers the potential and type of adverse impact that a fire may have on specific factors, including occupancy profiles, asset profiles, business issues and environmental impact.

Note that the above model follows the theme introduced for the scenario fire risk assessment formula provided previously. The best way to illustrate this is by using the worked examples from the previous chapter.

Metro railway station

Scenarios derived from Objectives Setting:

Scenario 1: Train on fire arriving at the station with the objective of minimising downtime of operations.

Scenario 2: Equipment rooms at platform level experiencing a fire with the objective of minimising downtime of operations.

Scenario 3: Uncontrolled small fires at any part of the station with the objective of preventing such fires or minimising their impact.

Scenario 4: Uncontrolled fire at low levels (platforms) requiring manual firefighting may lead to local environmental contamination of areas, including subterranean rivers.

Scenarios derived from Threat Analysis:

Scenario 1: Arson on a train leading to a "train on fire" in a running tunnel or entering a station.

Scenario 2: Arson in passenger walkways and at platform level.

Scenario 3: Fire in sub-surface passenger routes and platform leading to passenger confusion and delayed evacuation times.

Now, all scenarios should be assessed by the three-point cluster analysis model. This is illustrated in Table 9.

The next step is to analyse commonalities to provide a post cluster evaluation (Figure 16), reducing the assessment to a specific number of explicit scenarios.

One suggested method is to start with the impact cluster and then work back to the root cause of the fire scenario. From the analysis, it is possible to hone the scenarios into a fewer number. Note that it may be appropriate to reword or redefine the limiting conditions.

Table 9: Metro station: Cluster analysis

Ref.	Scenario	Ignition Cluster	Growth Cluster	Impact Cluster
1	Train on fire arriving at the station to minimise downtime of operations.	Fire on train	Onto platforms and in tunnels	Downtime of operations
2	Equipment rooms at platform level experiencing a fire with the objective of minimising downtime of operations	Platform level equipment room fires	Onto platforms and in tunnels	Downtime of operations
3	Uncontrolled small fires at any part of the station with the objective of preventing such fires or minimising their impact.	Small fires in any part of the station.	Fire growth in passenger walkways and platforms, fire growth in non-passenger areas.	To restrict the numbers of fires in terms of frequency and impact
4	Uncontrolled fire at platforms level, requiring manual firefighting, which may lead to local environmental contamination of area including subterranean rivers.	Platform level fires	Growth to a significant fire at platform level requiring professional firefighting	Environmental contamination of local area and subterranean river due to firefighting water runoff.
5	Arson on the train leading to train on fire in running tunnel or entering a station.	Fire on the train or in running tunnel	Onto platforms and in tunnels	Downtime of operations, impact on evacuation
6	Arson in passenger walkways and on the platform level.	Passenger walkways/ platforms	Throughout the passenger walkways and platforms	Untenable evacuation conditions
7	Fire in sub-surface passenger routes and platform leading to passenger confusion and delayed evacuation times.	Passenger walkways / platforms	Throughout the passenger walkways and platforms	Delayed evacuation times due to passenger confusion

Typically, the number of "impact clusters" will dictate the number of scenarios for further analysis. In this case, we have refined the number from seven scenarios to four:

Clustered Scenario 1: Consider operational downtime of the rail network due to fire growth in running tunnels and on platforms, caused by a fire on a train, or a platform, passenger walkway, or due to a small fire within a station.

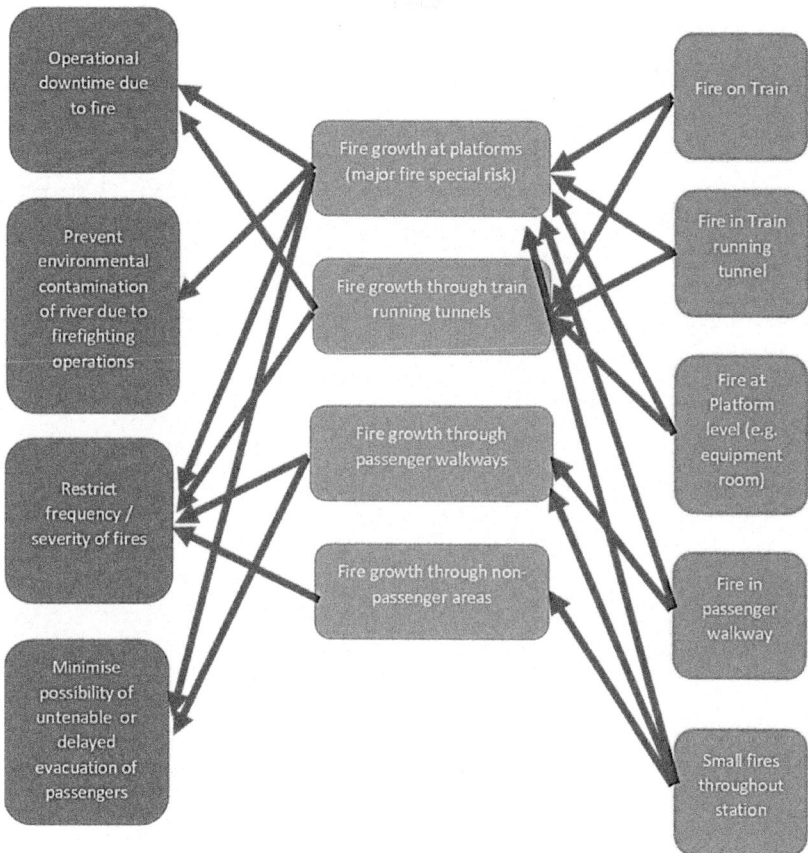

Figure 18: Cluster analysis of scenario fire risk (metro station)

Clustered Scenario 2: Prevent environmental contamination of a local subterranean river due to firefighting runoff caused by fire growth at platform level due to ignition at platform level or in a running tunnel.

Clustered Scenario 3: Minimize frequency and impact of small fires anywhere on the station premises.

Clustered Scenario 4: Optimise evacuation conditions (tenability and time to evacuation) due to fire originating and spreading at platform level and in passenger walkways.

These four clustered scenarios will then be assessed for scenario fire risk. This is shown later in this chapter.

School building.

Scenarios derived from Objectives Setting:

Scenario 1: A fire that could create untenable conditions for evacuation from sub-basement and roof areas.

Scenario 2: A fire on the 4th and 5th floor requiring firefighter assistance will currently expose them to the direct impact of a fire.

Scenario 3: A fire affecting the new computer centre that could lead to asset loss.

Scenario 4: A fire of sufficient severity to degrade the asbestos ceiling tiles and causing the release of asbestos fibres.

Scenarios derived from Threat Analysis:

Scenario 1: Arson by lighting waste bins located next to the school building.

Scenario 2: Failure of existing manual fire alarm system due to lack of maintenance.

Scenario 3: Uncontrolled fire in chemistry lab.

As with the railway station, all scenarios should be assessed by the three-point cluster analysis model (Table 10).

The next step is to analyse commonalities in the above to provide a post cluster evaluation to reduce the assessment to a specific number of explicit scenarios. The cluster interlinks are illustrated in Figure 17.

When compared to the example for the metro station, it is noticeable that cluster analysis has failed to reduce the number of fire scenarios. The main reason for this is that each of the chosen scenarios is not interdependent in any way. There is also a scenario included that is not strictly a fire scenario for additional determination, although it may have valid reasons for inclusion on the list. This is fire scenario number 6. In this case, that there are no directly relevant ignition and growth clusters. It is a fire safety management issue covering a failure to maintain fire protection systems.

We can now derive the fire scenarios following the cluster analysis. By undertaking the analysis, we have an opportunity to redefine and hone the scenario, and in some cases, expand the scenario if it originally covered more than one area for analysis. There may also not be the need to define both ignition and fire spread clusters separately if it is evident that they affect the same areas.

The revised clustered scenarios are as follows:

Clustered scenario 1: Untenable conditions for evacuation from the sub-basement area due to smoke and fire spreading into the sub-basement and impacting the escape route.

Table 10: School building: Cluster analysis

Ref.	Scenario	Ignition Cluster	Growth Cluster	Impact Cluster
1	A fire that could create untenable conditions for evacuation from sub-basement and roof areas.	Fire ignition in or near sub-basement / roof areas	Fire spread affecting sub-basement and roof areas through to protected escape routes	Untenable evacuation conditions in sub-basement and roof areas
2	A fire on the 4th and 5th floor requiring firefighter assistance without exposing them to the direct impact of a fire.	Fire ignition anywhere on the 4th or 5th floor.	Fire spread across 4th and 5th floor levels and into firefighter staircases.	Direct exposure of firefighters to effects of smoke and fire
3	A fire igniting or directly impacting the new computer centre that could lead to asset loss.	Fire igniting in the computer centre or in an area that could spread smoke into the computer centre.	Fire and smoke spread affecting computer centre	Asset loss of computer centre
4	A fire of sufficient severity to cause degradation of the asbestos ceiling tiles and causing the release of asbestos fibres	Fire ignition near areas of asbestos ceiling tiles.	Sufficient heat from fire spreading to asbestos ceiling tiles	Release of asbestos fibres die to heat degradation of asbestos ceiling tiles
5	Arson effort by lighting waste bins located next to the school building.	Fire igniting in an external waste bin	Spread of radiative heat from the external fire into the school building.	Loss of part or all of the school due to fire in a waste bin
6	Failure of existing manual fire alarm system due to lack of maintenance.	N/A	N/A	Failure of manual fire alarm system to operate due to lack of maintenance.
7	Uncontrolled fire in the chemistry lab.	Fire ignition caused by chemical equipment and processes.	Fire and explosion risk spread throughout the chemistry lab	Risk to both life and assets due to fire in the chemistry lab

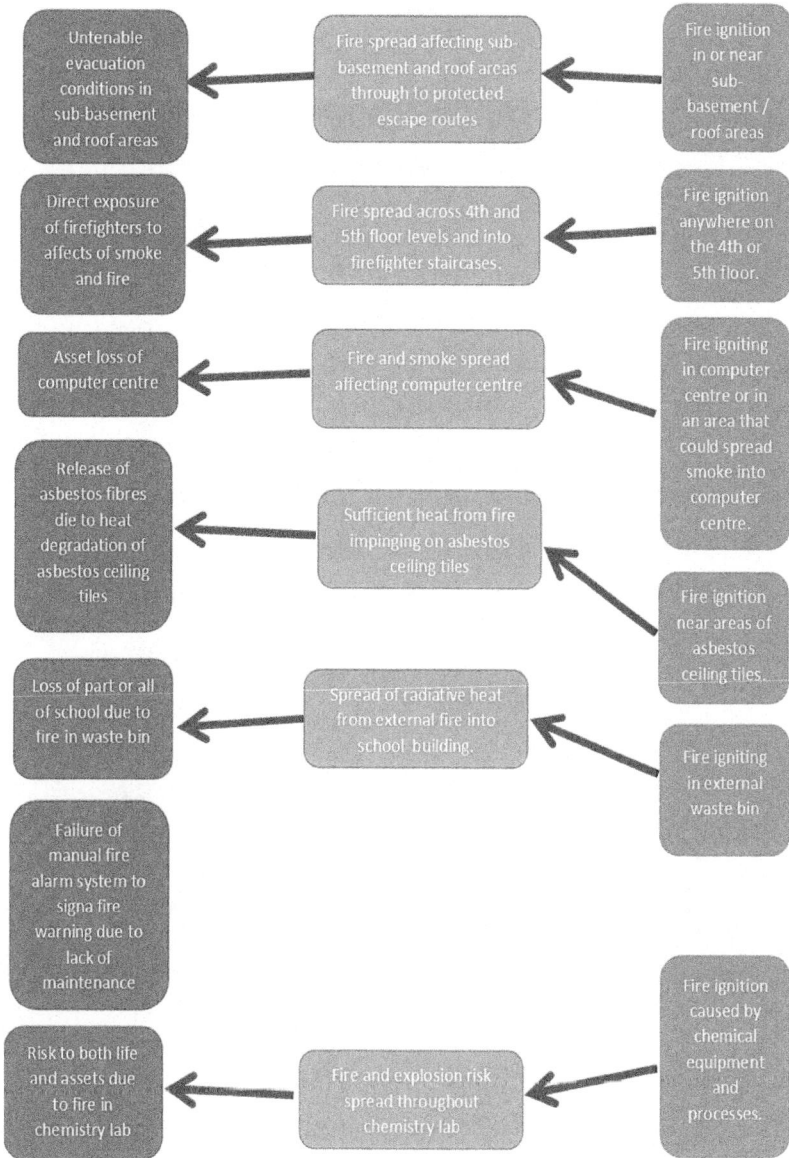

Figure 19: Cluster analysis of scenario fire risk (school building)

Clustered scenario 2: Untenable conditions for evacuation from the roof area due to smoke and fire spreading into the roof space and impacting the escape route.

Clustered scenario 3: Direct exposure of firefighters to the effects of fire and smoke if fighting a fire on the 4th and 5th floor levels.

Clustered scenario 4: Asset loss of computer centre due to smoke and fire caused by fire ignition in the centre or directly affecting it.

Clustered scenario 5 The release of asbestos fibres from ceiling tiles due to the direct impact of heat from a fire in those areas where the tiles are installed.

Clustered scenario 6: Fire damage to school building due to external waste bin fire impinging on the building.

Clustered scenario 7: Life threat and asset loss/damage in the chemistry lab due to fire ignition from the use of chemical equipment and processes.

Scenario Fire Risk (SFR) Analysis

Chapter 8 describes a process for assessing fire scenarios based on the probability of fire ignition and fire spread and the consequences of fire spread. As previously described, the singular version of the formula (Eq.8.7), where a scoring system for 0 to 5 is used, is:

$$SFR = P_g C_g \frac{(P_{ie}+P_{ip})}{10} \qquad \text{Eq.8.7}$$

Where SFR = Scenario Fire Risk, P_g = Probability of fire growth, and C_g = The consequences of fire growth, P_{ie} is the probability of ignition due to the environment hazard, and P_{ip} is the probability of ignition due to the process hazard.

Following the cluster analysis, the same method can be applied to evaluate the clustered fire scenarios.

We can utilise the same worked examples as used in this chapter to show the process.

Metro railway station

The clustered scenarios, detailed from the earlier analysis, are as follows:

1: Consider operational downtime of the rail network due to fire growth in running tunnels and on platforms, caused by a fire on a train, platform, passenger walkway, or due to a small fire within a station.

2: Prevent environmental contamination of a local subterranean river due to firefighting runoff caused by fire growth at the platform level due to ignition at this level or in a running tunnel.

3: Minimise frequency and impact of small fires anywhere on the station premises.

4: Optimise evacuation conditions (tenability and time to evacuation) due to a fire and smoke originating and spreading at platform level and in passenger walkways.

The *SFR* for each scenario can be calculated as shown in Table 11.

It is best if a group rather than an individual undertakes the task to reduce levels of subjectivity, given that the task itself is mainly subjective. This is also an occasion to make use of any statistical or performance data to qualify the results. The consequences of a fire are included in the table.

The analysis provides a comparative assessment methodology. Thus, stakeholders may, for example, set a benchmark that, for instance, all scenarios over a certain number should be evaluated further in order to determine suitable control measures.

Alternatively, the highest numbers may be chosen for further treatment. In this example, we will assume that scenario 2 has been chosen for further investigation.

Table 11: Metro station: SFR analysis

Scenario	C_g	P_{ie}	P_{ip}	P_g	SFR	Commentary
1	4	1	2	4	4.8	The environmental conditions for ignition are low, but the process ignition risk is rated slightly higher. Growth and spread are likely if the fire gets to a specific size.
2	5	1	2	4	6	The ignition and growth phases will reflect those for scenario 1, although the consequence rating is higher.
3	4	1	2	2	2.4	Small fires are infrequent and usually the result of mishandled processes, but the existing provisions usually contain fire growth.
4	5	1	1	4	4	The ignition environment and growth phases will reflect those for scenarios 1and 2. However, the process ignition rating is slightly lower when platforms and passageways are considered on their own, although the consequence rating is higher.

School building.

The scenarios, detailed from the earlier analysis, are as follows:

1: Untenable conditions for evacuation from the sub-basement area due to smoke and fire spreading into the sub-basement and impacting the escape route.

2: Untenable conditions for evacuation from the roof area due to smoke and fire spreading into the roof space and impacting the escape route.

3: Direct exposure of firefighters to the effects of fire and smoke if fighting a fire on the 4th and 5th floor levels.

4: Asset loss of computer centre due to smoke and fire caused by fire ignition in the centre or directly affecting it.

5: The release of asbestos fibres from ceiling tiles due to the direct impact of heat from a fire in those areas where the tiles are utilised.

6: Fire damage to school building due to external waste bin fire impinging on the building.

7: Life threat and asset loss/damage in chemistry lab due to fire ignition from use of chemical equipment and processes.

As with the metro railway station, Table 12 is completed as shown.

Table 12: School building: SFR analysis

Scenario	C_g	P_{ie}	P_{ip}	P_g	SFR	Commentary
1	4	1	3	2	3.2	The sub-basement environmental conditions for ignition are low, but the process ignition risk is high due to the abundance of plant. Fire and smoke growth and spread are contained by fire-resisting construction.
2	4	3	1	4	6.4	The roof environmental conditions for ignition are considered medium due to levels of exposed softwood and illicit storage of materials. The process ignition risk is lower due to the absence of any plant. Fire and smoke growth and spread are not contained as the roof is open to the lower floors due to non-fire resisting access hatches.
3	5	1	4	4	10	There is no substantial separation from the two staircases used by firefighters to the open plan risk areas on the 4th and 5th floor levels. This would have been a code requirement for modern buildings but was not required when the building was originally constructed. The environmental risk of fire is low, but these higher floors are used for physics and biological experimentation.
4	4	3	3	2	4.8	The computer centre had been fitted with an aspirating smoke detection system connecting in with the manual alarm system so there will be an early

Scenario	C_g	P_{ie}	P_{ip}	P_g	SFR	Commentary
						warning of a fire. This should reduce the growth of fire and smoke. The environmental ignition risk is not low given the abundance of power supply leads to power all the computers.
5	5	1	1	2	2	Asbestos ceiling tiles are confined to the common spaces where the environmental and process risk of ignition is considered low. Furthermore, it was noted that heat growth would need to be substantial (over 800C) to release fibres.
6	3	1	1	1	0.6	Waste bins had recently been relocated to a proprietary bin store.
7	4	2	4	1	3.2	The process risk of ignition in the chemistry lab is high, and the environmental risk is not considered insubstantial. Fire spread outside of the chemistry lab is controlled due to the compartmentation of the classroom.

From the above, scenario 3 is an obvious choice for further investigation. This is followed by scenario 2. Both choices will be considered in the next section.

Control measures

Once one or more scenarios have been chosen, there should be a method of determining the most suitable control measures. There may be multiple options that could satisfy a single scenario. Conversely, there may be a single control measure that will adequately satisfy more than one scenario. This will not be known until the evaluation has been completed.

BS PAS 911 introduced a methodology for comparing alternative options for fire safety and protection where multiple control measures are proposed. This is illustrated in Table 13. The idea is that each option is scored on each of three separate factors. The resulting three scores are then multiplied to derive a numeric value for each option. The higher the number, the most appropriate the option is. The three factors (evaluation criteria) recommended for comparison are:

Technical performance: The ability of a type of system or process to reliably perform its function. A high score deems that this is a highly effective measure.

Logistics: Features of the building and its occupancy that may restrict the use and maintenance of the chosen option. A high score deems that this is relatively and logistically easy to achieve.

Economic considerations: The option is commercially viable immediately and for its lifetime. A high score deems that this is economically viable.

Any number of options can be evaluated in this way. However, it is recommended that there is a weighting system applied to the evaluation to allow a choice of what is considered to be the most crucial measure – on a case by case basis. For instance, if logistics is thought to be the most critical measurement factor, a maximum score of 10 could be applied. For the least important, let us say economics, a maximum score of 5 could be used. In this way, the importance of the factor will skew the result to consider the relative priority of the three factors. As stated above, the number of options may vary from two to as many options as deemed relevant by those undertaking the assessment.

Table 13: Control measure options: Assessment table

Criteria	Option A	Option B	Option C
Performance (out of x)	N1 out of x	N2 out of x	N3 out of x
	x	x	x
Logistics (out of y)	N4 out of y	N5 out of y	N6 out of y
	x	x	x
Economics (out of z)	N7 out of z	N8 out of z	N9 out of z
Total score:	=	=	=

Worked Examples

To fully understand how this method can be used, I will take the two examples we have worked on to date. The scenario chosen for the metro underground station was:

Prevent environmental contamination of a local subterranean river due to firefighting runoff caused by fire growth at platform level due to ignition at platform level or in a running tunnel. This could be due to a platform level fire or a train on fire.

Several control measures could be applied involving prevention and protection measures or even a combination of measures. Let us assume that relevant stakeholders held a meeting, and the following suggestions were given:

Option A: Prevent or severely limit the possibility of a large fire at this level. Given that extensive passive and active fire protection levels are already used to protect risk rooms at the platform level, the focus is on a train on fire. Therefore, it was deduced that the only conceivable way to do this is to fit fire detection and fire control systems into all railway carriages. It was agreed that performance-wise, this would work. However, the logistics and economics would be such as this should be scored low.

Option B: Prevent firefighting runoff water from entering the subterranean riverbed by water-resisting construction between the station and running tunnels and the riverbed. This would fully achieve the objective although, it would provide extreme logistical problems and would be very expensive to achieve.

Option C: To provide trained first response teams to act on a fire scenario before it grows to a stage where professional firefighting is required. Given that all staff are trained in basic firefighting, this would require additional firefighting training for key staff and additional firefighting facilities at platform level to allow the tackling of a train fire.

When scoring, it was decided that performance should take major priority at a value of 10, with logistics and economics taking a lower priority at 5 each. It could be assumed that the stakeholders scored as shown in table 14.

Table 14: Metro station: Scenario control measures - evaluation of options

Criteria	Option A	Option B	Option C
Performance (out of 10)	7	9	7
	x	x	x
Logistics (out of 5)	2	1	4
	x	x	x
Economics (out of 5)	2	1	3
Total score:	**28**	**9**	**84**

Clearly from the scoring, option C is by far the preferred option.

Moving forward to the school building, the two chosen scenarios will need to be separately evaluated:

Scenario 1 (school): Untenable conditions for evacuation from the roof area due to smoke and fire spreading into the roof space and impacting the escape route.

It could be assumed that the following options for control measures were considered.

Option A: Upgrade passive fire protection for roof area separating risk areas from access areas through to the point of designated relative safety. In this case, it is realised that this would require improving fire compartmentation from the roof space area to the staircase and then protecting the staircase to the ground floor level due to the lack of protected staircases. With limited budgets, this may be too expensive and may require the school's closing for such an upgrade. It is realised that the option would also improve safety for other areas of the school building. Therefore, the performance will be scored highly.

Option B: Provide a fire watch service in the school building when anybody occupies the roof space. Although this will not enhance the fire strategy, it will require additional procedures. Not a costly or logistically difficult option. Its effectiveness is questionable.

Option C: Providing automatic fire detection to monitor the roof space and all risk areas that could impact the evacuation route to a place of safety. Given that the school already is fitted with a central manual alarm system, adding detection circuits will not be that difficult. It will provide additional assurance for those in the roof space providing additional evacuation time.

The school stakeholders determine that performance and logistics should be scored out of 7, with economics slightly higher at 8 because of limited budgets.

Table 15: School building: Scenario control measures - evaluation of options (1)

Criteria	Option A	Option B	Option C
Performance (out of 7)	6	4	6
	x	x	x
Logistics (out of 7)	3	6	4
	x	x	x
Economics (out of 8)	2	7	5
Total score:	36	168	120

From the analysis, option B is the preferred choice. This would require inclusion in the fire safety management strategy rather than necessitate physical fire protection.

Let us now move to the second scenario for the school building.

Scenario 2 (school): Direct exposure of firefighters to the effects of fire and smoke if fighting a fire on the 4th and 5th floor levels.

It could be assumed that the following options for control measures were considered.

Option A: Upgrade passive fire protection to provide a protected staircase and lobby to separate firefighters from risk areas: This is the most obvious option and could have been combined with the previous Option A to provide a complete solution. As with the previous example, this would require improving fire compartmentation from the 5th floor to the staircase and then protecting the staircase to ground floor level. This may be too expensive and may require the closing of the school for such an upgrade. The performance will be scored highly.

Option B: Provide an external staircase that leads directly into the upper levels for use by firefighters: This is a feasible option given the availability of land around the school building. The entry points into the building may not give direct access to the risk areas.

Option C: Rely on high reach fire appliances to access the school at the higher floor areas: The school's hardstanding allows for access around the perimeter. The local fire and rescue service do not currently have such appliances, so the logistics are, in the short term, effectively non-existent. There would, of course, be no cost to the school.

As before, the weightings remain the same.

Table 16: School building: Scenario control measures - evaluation of options (2)

Criteria	Option A	Option B	Option C
Performance (out of 7)	6	5	6
	x	x	x
Logistics (out of 7)	3	3	1
	x	x	x
Economics (out of 8)	2	4	8
Total score:	36	60	48

Option B is the preferred option from the analysis, although the score differential is less than for the other example. It may therefore be deemed appropriate for a second review.

In summary, the purpose and scope of many fire strategies are to ensure that a building or other form of infrastructure is compliant with national regulations, codes, and standards. Given that these documents concentrate on occupants' life safety, an opportunity to take a more holistic approach is often missed. There is a need to encourage fire engineers to think about the broader issues at a sufficiently early stage in any construction project to ensure that the resulting fire strategy is truly holistic.

Objective setting: Acknowledging that most existing fire safety codes and regulations concentrate on the life safety of occupants, it is postulated that a holistic approach needs to consider much more. This is based on the idea that there are four main objective groups relevant to every fire strategy: life safety, property protection, protection of business and its continuity, and protection of the environment. Each of these prime objectives is divided into secondary sub-objectives.

Threat analysis: Many strategy documents explicitly state that "this fire strategy does not cover extreme events". They rarely account for specific threats that may require a review of the provisions already contained within a fire strategy. A document referred to within this chapter identified six specific threat groups for separate analysis.

This chapter, and the previous chapters, provide a suggested methodology for the consideration of both objectives and threats. The assessment concludes with a list of fire scenarios that may not be explicitly covered in the national regulations and standards. These scenarios are then exposed to a series of tests, including scenario cluster analysis and scenario fire risk analysis, to derive a reduced and qualified set of fire scenarios. From this, additional suitable control measures can be applied.

Given that there may be several alternative control measures to cater for a single fire scenario, a method of evaluating and scoring these control measures is given.

The techniques proposed may appear to be laborious. However, the various steps could quite easily be incorporated into a computer algorithm. The only input that may be required will be in the completion of a questionnaire.

This could then translate a qualitative set of answers into quantitative analysis, with the answers quickly delivered to the stakeholders involved. The next chapter reviews another example of an original idea promoted in BS PAS 911 and converted into a semi-quantitative tool. I should point out that I cannot take all the credit because the methodology was first suggested to me by a professor of environmental engineering based in Lodz, Poland.

Paul Bryant

12 A fire strategy indexing system

Without comparative measurement, there can be no validation.

The development of a fire strategy can be lengthy, especially for complex building arrangements. It may require detailed evaluation, a thorough understanding of relevant codes, detailed assumptions, agreement on performance criteria and the possible use of fire and evacuation modelling. Sometimes the level of detail required may hide the overall intentions of the fire strategy - as discussed in previous chapters.

At the beginning of this book, I highlighted business models developed to "picture" the competitive nature of businesses within the sphere they operate. This simple expression provides a much easier method to appraise the business without delving into detailed reports instantly. In the same way, it is believed that a pictorial representation of a fire strategy can help deliver the central message of the strategy without the necessity of great levels of detail. This will provide a better understanding to all stakeholders, particularly those who are not professional fire engineers.

A diagram titled "the fire strategy value grid" was initially published in British Standard PAS 911 and is described below. In addition, there was a second iteration of my idea using a semiquantitative approach which I will explain later.

The idea of the diagram was to allow a quick and easy way of "picturing" a fire strategy. The picture should be apparent using one side of A4 paper. This method is captured in Figure 18 and designed to identify the main elements of a fire strategy. It is intended to allow the user of the diagram to show how they believe each element contributes to the overall fire strategy. The diagram allowed identification of the relative value of each of the eight elements appropriate for every fire strategy.

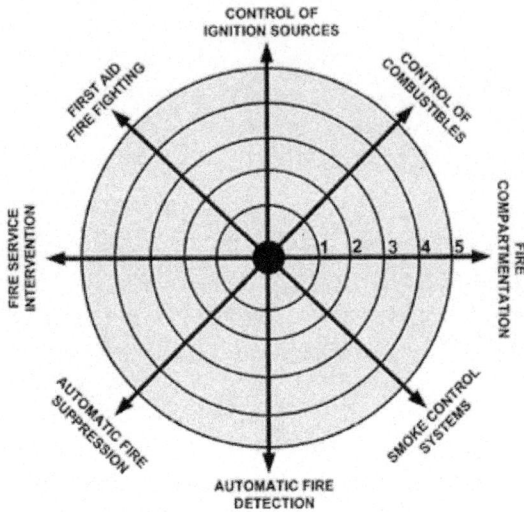

Figure 20: Fire strategy value grid (BS PAS 911)

As well as allowing the visualization of a fire strategy, it also allows for value analysis as the strategy develops. The eight factors are described below:

1. **Control of ignition sources**. *Issues for consideration will include an assessment of the location and use of electrical equipment, internal processes, and potential heat paths. Separating areas or rooms with highly ignitable sources from combustible materials can be highly effective.*

2. **Control of combustibles**. *Combustible materials may exist in the external and internal fabric of the*

building, such as exterior cladding and insulation materials, wall and ceiling linings, or in the fixtures and fittings. In addition, consideration should be given to the use of the buildings in terms of the storage of combustible materials.

3. **Fire compartmentation**. *Rather than thinking of fire compartmentation in absolute terms (e.g., 30- or 60-minute fire resistance ratings), we should think in terms of objectives. It is a fundamental objective to separate fire from occupants for as long as necessary for those occupants to escape to a place of safety. This generally translates into maintaining fire resistance partitions between the fire and persons for at least 30 minutes.*

Another objective is to allow firefighters to arrive at a building, enter the appropriate floor level, establish firefighting water supplies, and commence fighting the fire until it is fully extinguished. This may require up to a 120-minute separation between the firefighting access staircase/lift and the risk area, possibly, separated by a firefighting lobby of a rating of 60-minutes from the risk area.

Other objectives may be to allow parts of the building to continue to be used until a fire can be extinguished, or to require one part of the building to be substantially divided so that even an intense and prolonged fire in one part is highly unlikely to penetrate other parts. The "stay put" strategy for high rise residential buildings comes to mind here.

Note also, to assure a passive fire strategy, additional elements may need to be considered, such as the use of cavity barriers to protect unseen cavities in parts of a building.

Once the true purpose of fire compartmentation is understood, then the specification can be made that much easier. Remember that fire compartmentation is only as good as its weakest link.

4. **Smoke control systems**. *As with fire compartmentation, smoke control aims to provide for similar objectives, i.e., to protect the escape routes and allow for firefighting (in this case, manual firefighting). The simplest method of controlling smoke may be by opening a window or a vent. In most cases, and with many modern buildings, this will not be enough. Smoke control systems for life safety purposes will either keep escape routes free from smoke by keeping the smoke layer above head height or creating pressure differentials to keep specifically targeted areas free from smoke. Proprietary smoke control also allows firefighters to continue their operations without being exposed to dense smoke plumes. Basements are a good example. Note that smoke control may also be used to divert smoke from damaging high-value fixtures, equipment, and processes.*

5. **Automatic fire detection**. *Automatic fire detection is often considered a cornerstone of any fire strategy, but this need not always be the case. The first premise is that fire detection systems, on their own, do not actually do anything. Fire detection systems are, in effect, fire-monitoring systems. It is what you do with the signal once a fire has been detected that is relevant. It could be to initiate warning systems and control fire protection systems. Confidence in the system has also been proved to be an essential factor. A key requirement is that systems can discriminate between fire and non-fire or unwanted phenomena.*

6. **Automatic fire suppression**. *Fire suppression systems, such as sprinkler systems, may be chosen because third parties such as insurers specify them, or compliance with national law requires them, for instance, for buildings over a certain height. They may also be used as a "trade-off" to allow increased travel distances or reductions in the specification of fire separations. Those in the industry understand how successful sprinklers are in maintaining building fire safety and*

would benefit all building types, heights, and profiles. Nevertheless, some countries are more "sprinkler friendly" than others.

Sprinkler systems are only one example of automatically operating suppression systems to control a fire. There are many other types, from water-based options including fog/mist systems, deluge systems, foam systems and other specialist systems. There is a range of gas extinguishing systems, air volume inerting systems, powder-based systems, and systems that react with flame's free radicals. New ideas in fire suppression methods and techniques tend to come along frequently – all with specific benefits. We need to divide our thinking of fire suppression into total flooding systems - those designed to protect whole areas of the building, and local application systems – those used to suppress or extinguish fire directly at the source. It is the latter category where I see a promising future.

7. **Fire service intervention**. *Fire strategies may assume that the professional fire and rescue services will attend quickly and will tackle a fire anywhere in the building. This assumption relies on several factors. How will the arrival of firefighters, and the fire engines, be appropriately accommodated? How will they be met? How will they access the building, including upper and basement areas? Furthermore, the firefighting infrastructure must be available for firefighters to utilise. In some cases where professional levels of emergency response are essential, and the size and complexity of the building(s) warrants it, it may be appropriate to set up an in-house professional firefighting team.*

8. **First aid firefighting**. *Legislation and guidance have usually mandated that first aid firefighting equipment be provided practically everywhere outside domestic dwellings. However, many fire engineers and other parties do not treat first aid firefighting as a serious*

fire strategy measure. Some have questioned my inclusion of portable fire extinguishers as a critical part of a fire strategy. In 2002, the UK's Fire Extinguishing Trades Association [87] reviewed over 2,100 fire incidents. They found that in approximately 80% of the cases, a portable fire extinguisher successfully extinguished the fire, and in 75% of those cases, the fire department was not required to attend. In the US in 2010, portable fire extinguishers extinguished an estimated 5.32 million fires. It is much better to tackle the fire at the point of origin than to let it develop. First aid firefighting need not be regarded as a simplistic part of the fire strategy. I feel that this is an area where advances can and will be made.

The eight nodes represent primary strategic factors. Secondary factors such as alarm sounder systems, fire doors, extract systems, and fire prevention policies will follow from the primary factors. The idea of the diagram is to allow each of the eight factors to be separately evaluated and scored - from zero to five, based on their perceived relative importance to the strategy. The diagram was developed as a *first round* of analysis, although my book "Fire strategies - strategic thinking" highlights that its real benefit is in the regular revisiting of the strategy as its preparation progresses.

An alternative and novel use of the diagram was first proposed in Poland in 2017 and published in a polish language book in 2018[88]. This derivative concept took the same diagram but amended it in the following ways:

a) Increased the maximum scoring from 5 to 25.

b) Introduced a series of questions for each node to allow a more focused and less subjective evaluation.

c) Introduced the concept of comparing a baseline fire strategy (based upon national codes) with the actual fire strategy developed for the building either under construction or retrospectively for existing buildings.

d) Developed a single scoring factor for fire strategies.

The Polish assessment technique used a novel risk assessment methodology and was first presented in a scientific journal article[89]. The method described in the Polish variation included the use of fire engineering analysis and computational fluid dynamic (CFD) modelling to demonstrate the efficacy of the fire protection, both proposed or existing, against a baseline fire strategy. The analysis enabled the evaluation of the following parameters:

a) Fire and smoke development within a room or area.

b) The potential response time of fire detection systems.

c) The impact on the tenability of evacuation routes by smoke and heat development, with or without smoke control systems.

d) The effect of fire suppression systems on the control of heat and smoke.

e) The effect of fire compartmentation on smoke and heat movement.

A team of engineering specialist stakeholders should ideally undertake the fire strategy evaluation. For each evaluated building (or part, e.g., a fire zone), the methodology compares two fire strategies: the baseline strategy (default, based on the building risk profile or determined individually) and the actual strategy (real, realised for a new build project or existing building). The level (relevance) of each of the eight fire safety factors is scored from zero to twenty-five for both the baseline and the actual fire strategies.

Formulating a questionnaire

The following questionnaire (Table 17) is a derivative of the format prepared and used in Poland. Both the baseline fire strategy and the actual fire strategy should be independently assessed using the same questionnaire.

The questionnaire incorporates a series of features that can be summed together to come to a final score. In some cases, there is an "or" option. Individual factors within a question may typically score up to 5 or 10. If it is believed that the factor is required (for baseline cases) or provided (for actual cases) but is not complete, then the feature may be scored between zero and the maximum score for that question.

If the building is designed precisely in accordance with the national standards, the baseline and actual fire strategies should match. In other cases, the evaluation can show where adjustments have been made, such as in the provision of a sprinkler system to allow for reductions in other elements.

Let us use the same building examples as before, a metro station and a school building. In the case of the metro station, the baseline rules would follow a specific code drawn up by the metro operator, which would typically vary slightly from the national codes. In the case of the school, a national code would be applicable. In the case of a school in the UK, BS 9999 can be adopted with a risk profile of A2 (a building where occupants know the escape routes and with the potential for a medium growth fire).

Using the scoring system of Table 17, I evaluated the baseline conditions for both building profiles. The results are provided in Table 18.

Table 17: Fire strategy grid questionnaire

Node	Questionnaire	Max Score
1 Control of ignition sources (CI)	Basic management documented controls (≤5) + regular testing of electrical systems (≤5) + restriction of all forms of ignition to protected and fire separated areas (≤10) + specialised ignition control methods (≤5).	25
2 Control of combust-ibles (CC)	Basic management documented controls (≤5) + strict controls on the specification of internal and external building fascias and linings (≤10) + strict controls on the specification of fixtures, fittings, and equipment (≤3) + restricted allowances on size, location or type of combustible materials in common spaces (≤5) + creation of sterile areas in escape routes by locating combustible materials in controlled and contained environments (≤2).	25
3 Fire compart-mentation (FC)	Requirements for the structural integrity of the building against fire (≤5) + fire separation of escape routes from risk areas (≤5) + enhanced fire separation to allow for firefighting operations (≤5) + enhanced fire separations to maintain specific occupied areas of the building until firefighting operations have been completed (≤5) + enhanced fire separations to allow for protection of designated assets (≤5). Note that, in all the above, consideration should be given to concealed areas such as cavities.	25
4 Smoke control systems (SC)	[Use of simple methods to vent critical areas, e.g., the opening of windows in escape stairs and lobbies (≤5), OR use of automatically operating proprietary venting arrangements (≤10)] + mechanical smoke extract or pressurization systems in escape staircases (≤5) + mechanical smoke extract systems to enable variations of other elements of the fire strategy (e.g. extended travel distances) (≤5) + use of smoke control systems to allow continuance of critical operations during a fire incident (≤5).	25
5 Fire detection (FD)	An automatic fire alarm system using manual activation only (≤5) + monitoring escape routes by fire detectors (≤5) + monitoring of hazard areas (≤5) + monitoring all remaining areas (≤5) + enhanced specialist monitoring methods (≤5).	25
6 Fire suppress-ion (FS)	Suppression systems to protect high hazard equipment (≤5) + systems to protect hazard rooms or areas (≤5) + systems to protect critical parts of the building (≤5) + systems to protect all remaining areas of the building (≤10)	25
7 Fire service interven-tion (FI)	[Manual method only for contacting fire service (≤5) OR automatic or 24/7 trained manual contact (≤10)] + response time to be within national guidance (≤5) + proprietary facilities for access to all parts of the building (≤5) + facilities for firefighting water supplies (≤5)	25
8 First aid firefighting (FA)	Provision of portable fire extinguishers (≤5) + personnel trained in their use (≤5) + specialist firefighting systems for specific risks (≤5) + trained professional firefighters on-site for rapid attendance to a fire (≤10)	25

Paul Bryant

Table 18: Fire strategy value grid- baseline conditions

Node	Metro Station	School Building
1 Control of ignition sources (CI)	20	15
2 Control of combustibles (CC)	25	10
3 Fire compartmentation (FC)	20	15
4 Smoke control systems (SC)	10	10
5 Fire detection (FD)	20	5
6 Fire suppression (FS)	15	0
7 Fire service intervention (FI)	25	20
8 First aid firefighting (FA)	15	5

Now let us assume that we have re-evaluated the actual fire safety factors in both cases. In the case of the metro station, the actual strategy was very close to the baseline strategy. The only variation was that portable fire extinguishers had been removed from the platform levels due to a high incidence of vandalism.

In the case of the school, it was found that for the control of combustibles, some of the linings specified were not fire-resisting. Non-resisting glazing was specified for incorporation into some of the fire resisting partitions. On the plus side, a sprinkler system had been fitted throughout. Table 19 provides the results.

Table 19: Fire strategy value grid- Baseline and Actual conditions

Node	Metro Station		School Building	
	B/L	Act	B/L	Act
1 Control of ignition sources (CI)	20	20	15	12
2 Control of combustibles (CC)	25	25	10	7
3 Fire Compartmentation (FC)	20	20	15	8
4 Smoke Control Systems (SC)	10	10	10	10
5 Fire Detection (FD)	20	20	5	5
6 Fire Suppression (FS)	15	15	0	25
7 Fire service intervention (FI)	25	25	20	20
8 First aid firefighting (FA)	15	12	5	5

The table can be converted to the strategy value diagrams as shown in Figures 19 and 20. The next section provides a method to evaluate the strategy and define a *fire strategy risk index.*

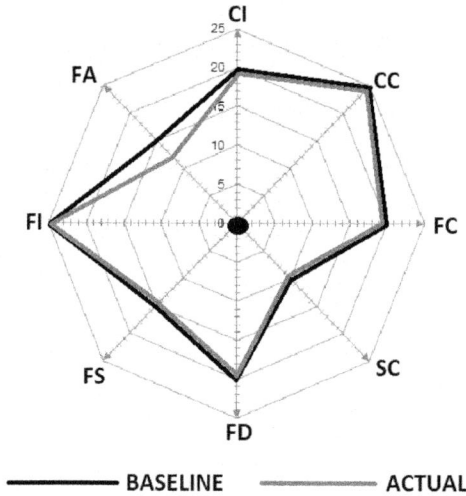

Figure 21: Strategy value grid - Metro station

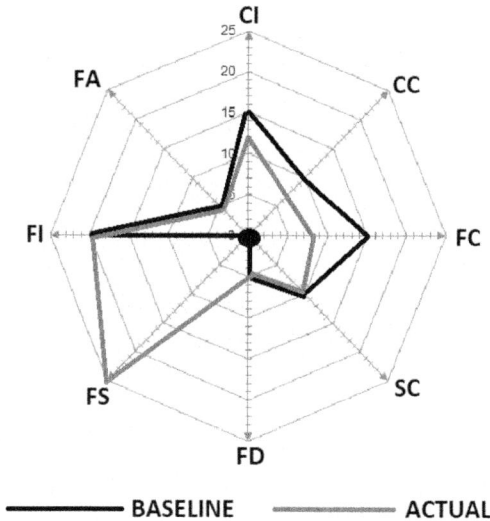

Figure 22: Strategy value grid – School building

Before we move into the semiquantitative analysis, let us first evaluate the diagrams themselves. The pattern can be quite revealing. The area within the pattern provides an idea of the resources and costs associated with the provisions required by the strategy. For instance, if each element is scored as twenty-five, the pattern will form the whole outer core of the diagram, with the maximum area taken up. Consequently, the fire safety and protection provisions are likely to be costly to implement and extremely resource hungry. Conversely, a shape with a much smaller footprint will be much more affordable. However, it must still be effective and answer all the issues from the earlier objectives setting.

The way the pattern sits on the diagram also can tell something about the type of strategy. For a pattern predominantly in the upper quadrant, the strategy will place greater reliance on fire safety management. A pattern towards the lower quadrant will place greater reliance on active fire protection. A shape on the left-hand side indicates that the strategy will rely on suppressing a fire. A shape to the right places greater reliance on containment and control of fire by structural means. An item scoring highly in the diagram is likely to be an essential feature of the final written strategy.

When reviewing the examples, the resource requirement for a metro station is understandably more intensive than for a school building. The metro station relies on high levels of safety management and systems with an emphasis on firefighting. The school building would typically rely on a reasonable level of fire safety management, the maintenance of some levels of fire containment, the intervention of the fire and rescue services, and less reliance on systems. The inclusion of sprinklers for the school has distorted the shape of the actual case.

Fire strategy risk index calculation

The last step of fire strategy evaluation is a Fire Strategy Risk Index (*FSRI*) calculation. This calculation aims to amalgamate each of the individual scores for both the baseline and actual conditions.

It is the Gretener Method[90], developed by Swiss engineer Max Gretener, that was chosen as a basis for the broader assessment of fire strategies. Dr Gretener focused on improvements in methods to assess property risk analysis for the insurance industry in the 1960s, and the concepts he developed were not dissimilar from my objectives. The method promotes the calculation of potential hazard and protective measures values used for the final fire hazard index calculation. Empirical figures are used, estimated individually for the building, and based upon the level of applied fire protection. It allows for comparison, with solutions either generic or required by national legislation. The notion follows most probability type risk assessments, where fire risk was assumed to be a product of hazard severity and loss expectation represented by the fire frequency of ignition and presented in Eq. 12.1.[91]

The purpose of the *Fire Strategy Risk Index* (*FSRI*) is to provide a score for both the baseline and actual fire strategies based upon the product of the *Fire Hazard Index* (*FHI*) and the *Frequency of Ignition* (*Fi*) of which are found by analysis and existing codes.

$$FSRI = FHI \cdot Fi \qquad \text{Eq. 12.1}$$

The *Frequency of Ignition* is one of the parameters of most probabilistic risk assessments and is covered explicitly in British Standard PD7974-7. It is usually the initiating event in most event trees and can be a base event in fault trees. For example, offices typically have a frequency of ignition per year of around 0.4×10^{-2}, whilst, for hotels and restaurants, this increases to

4.6x10^{-2}. For our cases, schools are rated at 1.4x10^{-2} and public buildings and services (which includes metro stations) at 1.8x10^{-2}.

The fire hazard severity, as represented by the *Fire Hazard Index* from Eq. 12.1, is proportional to the Potential Hazard (*PH*), reduced by the Protective Measures (*PM*) applied, as shown in Eq. 12.2.

$$FHI = \frac{PH}{PM} \cdot 100 \qquad \text{Eq. 12.2}$$

The original Gretener formula derived numerical factors for fire initiation and spread, with factors for fire protection. The idea used in the method presented here uses the values achieved from the scoring of each fire safety factor for the baseline and actual strategies. The *FHI* is clearly a ratio. The use of "100" is to balance the frequency of ignition typical values. Clearly, the lower the value of *FHI*, the better.

Introducing a weighting system

When creating a scoring index for a fire strategy using both baseline and actual conditions, it could be assumed that, before the evaluation, each of the nodes is equally important. The opportunity to allow stakeholders to adjust the criticality of each node makes sense if the process is meant to be adopted globally. Adjustments may be for the following reasons:

a) Requirements and guidance of national legislation and codes vary in focus for fire strategies.

a) The results of objectives setting and threat analysis favour some nodes more than others.

b) Building specific issues where one or more stakeholders deem is essential.

The weighting value (*W*) should be given to each of the eight nodes, with the total coming to unity.

Consequently, for the case of each node being equally valued, then the value of *W* for all nodes will be 0.125

A value for protective measures (PM) is obtained from the formula below (Eq. 12.3) by aggregating the points obtained from assessing each fire safety factor and adjusted by the appropriate weighting factors.

$$PM = (W_{CI} \cdot A_{CI}) + (W_{CC} \cdot A_{CC}) + (W_{FC} \cdot A_{FC}) + (W_{SC} \cdot A_{SC}) + (W_{FD} \cdot A_{FD}) + (W_{FS} \cdot A_{FS}) + (W_{FI} \cdot A_{FI}) + (W_{FA} \cdot A_{FA})$$

Eq. 12.3

The Potential Hazard (*PH*) can be determined by assuming that, for the case where baseline conditions are precisely met, it will be assumed that *PH* = *PM*; therefore, the *FHI* will be exactly unity. The *PH* will use the same weightings *W* but will apply them to the original baseline scoring *B*. Similarly,

$$PH = W_{CI} \cdot B_{CI} + W_{CC} \cdot B_{CC} + W_{FC} \cdot B_{FC} + W_{SC} \cdot B_{SC} + W_{FD} \cdot B_{FD} + W_{FS} \cdot B_{FS} + W_{FI} \cdot B_{FI} + W_{FA} \cdot B_{FA}$$

Eq. 12.4

Let us use the above for the metro station and school examples.

Metro station

In this case, it has been agreed that each node should be equally weighted. The scoring given in Table 19 earlier is included in Table 20 to help calculate the *PM*.

Table 20: PM calculation - metro station

Metro Station					
Node	Weighting	Baseline	Actual	PM (B/L)	PM (Act)
CI	0.125	20	20	2.5	2.5
CC	0.125	25	25	3.125	3.125
FC	0.125	20	20	2.5	2.5
SC	0.125	19	10	2.375	1.25
FD	0.125	20	20	2.5	2.5
FS	0.125	15	15	1.875	1.875
FI	0.125	25	25	3.125	3.125
FA	0.125	15	12	1.875	1.5
Total:	1			19.875	18.375

Baseline strategy: PM = 19.875

Actual strategy: PM = 18.375

PH (in both cases) = 19.875 / 100 = 0.199

Actual strategy: $FHI = PH/PM \times 100 = (0.199/18.375) \times 100 = 0.744$

(Note that for the baseline strategy: FHI = 1)

The value of Fi for public buildings is given as Fi of $1.8 \cdot 10^{-2}$ as provided by PD 7974-7.

Therefore, the Fire Strategy Risk Index (FSRI) for both strategies is:

Baseline strategy: $FSRI = FHI \cdot Fi = 1 \times 1.8 \times 10^{-2} = 1.8 \times 10^{-2}$

Actual strategy: $FSRI = FHI \cdot Fi = 1.08 \times 1.8 \times 10^{-2} = 1.94 \times 10^{-2}$

In this case, the $FSRI$ for the actual fire strategy is slightly higher than the baseline. The difference is due to the reduction of first aid fire extinguishing systems. The result concludes that the risk requires re-evaluation.

School building

For this example, it has been agreed that each node should *not* be equally weighted based upon a threat analysis and the increased risk of arson affecting schools within the local community. It had been suggested that there is increased importance for a sprinkler system and less for portable fire extinguishers, which are regularly vandalised and cannot be relied upon.

As with the metro station, the values of the baseline and actual scoring taken from Table 19 are used to calculate the *PM* for both. Table 21 provides the results.

Table 21: PM calculation – school building

	School Building				
Node	Weighting	Baseline	Actual	PM (B/L)	PM (Act)
CI	0.125	15	12	1.875	1.5
CC	0.125	10	7	1.25	0.875
FC	0.125	15	8	1.875	1
SC	0.125	10	10	1.25	1.25
FD	0.125	5	5	0.625	0.625
FS	0.2	0	25	0	5
FI	0.125	20	29	2.5	3.625
FA	0.05	5	5	0.25	0.25
Total:	1			**9.625**	**14.125**

Baseline strategy: $PM = 9.625$

Actual strategy: $PM = 14.125$

PH (in both cases) $= 9.625 / 100 = 0.096$

Actual strategy: $FHI = PH/PM \times 100 = (0.096/14.125) \times 100 = 0.679$

(Note that for the baseline strategy: $FHI = 1$)

The value of Fi for public buildings is given as *Fi* of $1.4 \cdot 10^{-2}$ as given by PD 7974-7.

Therefore, the Fire Strategy Risk Index (FSRI) for both strategies is:

Baseline strategy: $FSRI = FHI{\cdot}Fi = 1 \times 1.4 \text{x} 10^{-2} = 1.4 \text{x} 10^{-2}$

Actual strategy: $FSRI = FHI{\cdot}Fi = 0.679 \times 1.4 \text{x} 10^{-2} = 0.95 \text{ x} 10^{-2}$

In this case the *FSRI* for the actual strategy is far lower than the baseline, largely due to the fitting of a sprinkler system.

The inclusion of the frequency of ignition may not change the ratio when comparing actual and baseline strategies for any one building. The benefit is where different FSRI's between building profiles are compared. The frequency of ignition provides a statistical adjustment to enable like for like comparisons. I should point out that I do have concerns as to the values of the frequency of ignition. The figures given in British Standard PD7974-7 derive from valuations as far back as the 1970s. Buildings and their uses have changed in that time, so perhaps this method would need to derive a more relevant and building profile-specific set of figures.

In both the examples used, we evaluated a single actual fire strategy. Why not assess multiple "actual" examples. Alternative strategies can be investigated by using CFD modelling to assess the efficacy of adjusting parameters such as fire suppression and smoke control options.

Perhaps even the baseline assessment could be varied by comparing a purely prescriptive baseline with one that uses a performance-based approach. Quite possibly, a catalogue of fire strategy risk indices can be built up over time and used for comparison.

13 Stakeholder analysis

There may be more groups with a vested interest in firesafe buildings than may be apparent.

This book intends to provide a modified methodology of fire strategy formulation and evaluation. The term "holistic" was chosen to highlight that a fire strategy needs to take an all-encompassing and global approach. I have pointed out that a typical feature of many fire strategies is in the narrow focus of formulation. Fire engineers set out to ensure that a fire safety design solution is compliant with the relevant (life safety) codes that support national fire safety regulations. The process does not encourage a proper degree of "holistic" thinking about all possible issues at an early stage of a construction project.

One of the primary reasons behind this book is to improve the auditability of fire strategy formulation. The needs of those undertaking the audit and approval of the fire strategy must be adequately considered. The enforcement agencies or authorities are the primary stakeholders in the process. Similarly, those formulating the strategy, the fire engineers working for the building project team, will also be central to the development of a fire strategy. There are many other stakeholders with a greater or lesser degree of involvement.

To ensure that the holistic fire strategy methodology is available to all, it would make sense that it is accessible via a web-based portal that the project teams and enforcement authorities could access. The whole process could be managed online – avoiding the need for paper but allowing for paper versions if so wanted.

This chapter undertakes a stakeholder analysis to determine how best *Holistic Fire Strategies* can be used.

There are potentially numerous stakeholder groups involved in any building project, from the first feasibility study through to the daily operational uses of the building. Hence, the concept should consider if and how it can improve each stakeholder's lot. A typical list of stakeholders is given below. Their needs and requirements are discussed.

Enforcement authority

"Enforcement authority" is a term to cover fire authorities and civil defence groups, and agencies whose job is to approve fire strategies for new build projects or major renovations to existing buildings. The enforcement authority will want a straightforward, consistent method to evaluate fire engineered design solutions. The purpose of this is to save them from tackling the myriad of solution formats and formulation processes that they currently must deal with. The outcome should therefore improve resource efficiency in the approval process. The method may also include some initial verification, or third-party peer review, to provide the agency with additional confidence in the presented strategies.

Another practical aspect of approval is the receipt of paper-based fire strategies. Could there not be an alternative method when the strategies are prepared and submitted online? Perhaps the enforcement authority would like the ability to instantly view the progress for any building project at a time they wish? It is suggested that the online method could incorporate features for the approval process itself by allowing enforcement authorities to confirm acceptance of the strategy, to suggest amendments to the strategy, or simply to reject the strategy – all online and with a suitably resilient audit trail.

Building Construction Project Team

A key influencer in the success of Holistic Fire Strategies will be those who are currently responsible for ensuring that the building will be suitably compliant and that it will meet all relevant requirements. The building project management team will ensure that the fire strategy meets the various stakeholders' requirements. Their prime motivation will be to deliver the project within budget and on time. In addition, they will want to ensure that every aspect of the project is adequately controlled and that any risks along the way are minimised. One of the critical risks for any building project is the fear of a "trip up", and fire safety regulation has historically been an area where this can occur. So how could the Holistic Fire Strategies approach assist them?

Many Project Managers have put up with listening to some of their experts around the meeting table, especially those with opinions. *Unfortunately, opinion engineering* without the necessary support of knowledge and experience is re-appearing. It could potentially lead to project problems further down the line, especially if that opinion is shown to be wrong. By introducing a consistent methodology, any subjectivity should be limited. Furthermore, a system that provides an internationally accepted approach should limit the chances of project-based errors.

Profitability will also play a key factor. The project team will want any new methods to be no more expensive than existing methods. A method that vastly improves efficiency in terms of cost, time, and resource, and delivers an acceptable design, will be easily attractive.

Fire engineer

Even if a fire engineer is part of the project team or not, their requirements may differ from that team. From experience, many professional fire engineers do fear that they may make a mistake along the way. They are concerned that they may have made the wrong decisions. They may have applied the wrong code requirements or misinterpreted some of the more detailed aspects of the project.

Would they not greatly appreciate a system that assists with their design and provides them with greater assurance that the assumptions they make are correct? What if the method incorporated a form of peer review that confirmed that their decisions are correct? Furthermore, what if the method provides them with all relevant code data at hand, based upon the region in which the building is being constructed and its intended use?

A method that expedites the fire strategy approval process would be a godsend to many. How many times have fire engineers had to wait around for days, weeks, and even months to get a response from enforcement authorities?

Architect

Architects are increasingly seeking to break the boundaries when it comes to building design. This can cause friction between them and those who are responsible for fire safety. Moreover, let us not forget that the best architectural practices work globally rather than nationally. They find it highly frustrating that one set of fire safety rules applies in one country and another set applies in another, even if the building they are designing are almost identical.

The provision of a global method will improve the consistency of approach and less confusion.

Regional Authority

Regional authorities worldwide usually have a legislative, political, commercial, and ethical responsibility to ensure that a building constructed within their jurisdiction is *firesafe*. Typically, part of their responsibility is to ensure that proper safeguards are applied at every step of a building construction process. They may employ the services of building inspectors who will take an active interest in the project. They may also work closely with enforcement authorities and be part of the approval process.

They will value a method that improves the consistency of approach and an improved audit trail of decisions made. They will want to be adequately prepared in the event of a major fire in one of their portfolios of buildings. A method that assists with this will be welcomed.

Government

National governments are politically inclined towards best fire safety practices. One major fire with disastrous consequences could effectively lead to a loss of confidence in their leadership. Governments are responsible for fire safety legislation and will want to encourage concepts that will improve fire safety. They would also welcome initiatives that cut down bureaucracy and improve efficiency.

The Fire Industry

Any new concept or methodology will undoubtedly need to be accepted by the national, if not international, fire community. They will always be interested in solutions that make their life easier. Sadly, there will always be a negatively biased contingent who will not embrace new ideas readily. This has been borne out by experience; many new ideas and techniques tend to find hurdles along the way. It could be said that this is due to the need to take a risk-averse approach, given that the subject matter is predominantly life safety. There are also other theories, of course.

The Construction Industry

The construction industry is generally dynamic and constantly appraising new building techniques and materials. They also make up a large part of a country's economy. The development and take-up of Building Information Modelling (BIM) systems is one example of when the industry continues to break ground. Therefore, any new solution that improves the input, consistency, and audit trail for the fire safety of buildings may be welcomed.

Other Interested Parties

There will always be other parties who may be involved on the periphery or may take a more central role. Insurers are one such body who will be interested in ensuring that their requirements are implemented. The broader aspects of objectives setting, including property, business and environmental protection, may be factors that, to date, have not been included in the primary fire engineering design objectives.

Similarly, there will be vested interest groups with interests in one or more aspects of the building and its operations, all of which will have their own set of requirements that may need to be met. A holistic approach will incorporate all such requirements based on the type of analysis provided by the method.

The Building Owner / Occupier

This stakeholder is probably the most important of all, even if they are not directly involved in the building design, construction, and approval process concerning fire safety. They will naturally assume that all appropriate measures have been used in the development of the fire strategy. They will also expect that the needs of their specific requirements, including the occupancy and uses of the building, will have been appropriately considered. A system that takes away subjectivity, improves coordination, and provides a more holistic approach will more effectively cater to the building occupier's needs.

Stakeholder relationship with Holistic Fire Strategies

Figure 21 illustrates an "onion skin" diagram of how the concepts behind Holistic Fire Strategies will impact the various stakeholders. At the core is the concept itself. Directly surrounding the core will be the three stakeholders who will use the concept; the project team, the fire engineer, and the enforcement authority.

The next outward layer includes those affected by the concept but are not directly involved. This will include the building owner and occupiers, regional authorities, architects, and other interest groups such as insurers. Finally, the outer layer incorporates those not involved with the project but whose interests will be relevant to how the concept is utilised.

The Holistic Requirements Model is derived from a systems approach to the classification of system requirements. It provides a consistent analysis framework that can be used when interpreting, clarifying and deducing system requirements from a set of customer/stakeholder needs. The model was developed by Dr Stuart Burge [92].

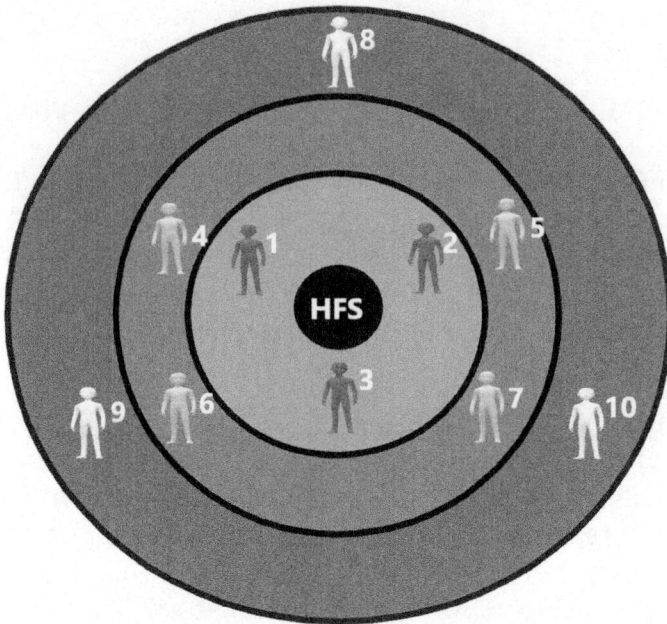

Figure 23: Stakeholder onion skin illustration (1 = enforcement authority, 2 = project team, 3 = fire engineer, 4 = architect, 5 = insurer, 6 = building owner, 7 = regulatory authority, 8 = fire industry, 9 = construction industry, 10 = government

The model is designed to apply a structured approach for the analysis of any type of system with the main components illustrated in Figure 22. Note that the term "holistic" is entirely coincidental with the subject of this book, but the relevance will be made clear.

The Holistic Requirements Model is named because it provides a complete and consistent model for classifying and organizing any set of requirements of a system. The model identifies three basic requirement types:

a) Operational Requirements.

b) Functional Requirements.

c) Non-Functional Requirements.

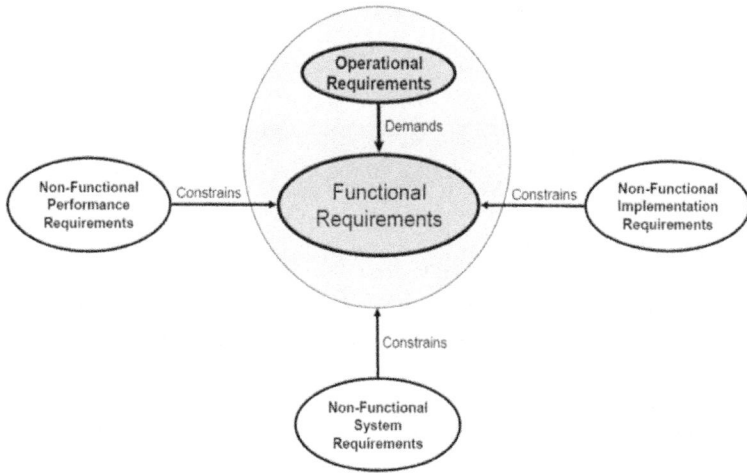

Figure 24: Holistic requirements model

Systems theory states that all systems must have a purpose and context. Furthermore, the prime system purpose can often be broken down into a few sub-purposes, sub-sub-purposes, and so on. These two systems theory facts lead to the definition of the operational and functional requirements of a system as:

Operational Requirements define the primary purpose of a system.

Functional Requirements specify what the system must do to achieve the Operational Requirements. Thus, they capture and define the sub-purposes of the overall system purpose (given in the operational requirements). Typically, functions use verbal phrases.

Non-Functional Requirements are effectively constraints on the system that define performance considerations.

Non-Functional Performance Requirements define how a particular feature must perform.

Non-Functional System Requirements define the constraints that affect the whole or a significant proportion of the system and include style, weight, reliability, cost, etc.

Non-Functional Implementation Requirements define how a system is to be built. These may be specific requirements from a potential user, or they may be legislative requirements.

The Holistic Requirements Model (HRM) functions could help decide how we can define the Holistic Fire Strategies system based upon the requirements of the various stakeholders identified earlier. Each stakeholder will need to be assessed to determine potential requirements for the system. Note that some requirements may be duplicated across two or more stakeholders. This is not a bad thing as this provides additional bolstering of some of the eventual functions of the system.

From that point on, we need to determine if each requirement type is:

- Operational (O)

- Functional (F)

- Non-Functional Performance (NFP)

- Non-Functional System (NFS)

- Non- Functional Implementation (NFI)

The assessment will help identify the primary dynamics of the Holistic Fire Strategies Model and the constraints. The main Operational Requirement will, in effect, be the topic of this book:

Operational Requirement: To provide a globally accessible method of formulating and evaluating fire strategies considering national requirements together with a wider consideration of objectives and threat analysis.

The term "holistic" is not included in the statement because it is a term used to describe the concept rather than how it should operate.

Next, the main functional requirements and non-functional constraints are to be defined on a stakeholder-by-stakeholder basis. In many of the requirements of different stakeholders, it will be seen that they are remarkably similar. The following illustrates the method. Table 22 provides stakeholder requirements.

Table 22: Stakeholder requirements

STAKEHOLDER	REQUIREMENT	HRM TYPE
1,2,3,4	The system should make use of a consistent framework. Ideally, all submitted fire strategies will follow the same format, making both preparation and auditing easier.	F
1,2,3,4,5,6, 7,8,9,10	The process should recognise national regulations and codes. There should be a built in method to allow fire strategies to follow national requirements and then amend as appropriate.	NFI
1,2,3,7,8	The auditing process for holistic fire strategies should be made as straightforward and unambiguous as possible.	NFP
2,3	The system should promote the best fire engineering knowledge. This could include a form of online code requirement search function based upon building profile and geographical region.	NFP
1,2,3,4	The system should require less time in assessing fire strategies than is currently required.	NFP
1,2,3,7	The system should provide an audit trail for decisions made: Automatic logging for decisions made by all parties.	NFS
1,2,3,8	Online rather than paper-based strategy review: The system should be accessible online. There should be an option to download paper-based strategies.	NFS
2,3,4,5,6	The concept should ensure that all owner/occupier needs are appropriately met. In addition, the system should ensure that all stakeholder objectives and concerns can be fed into the methodology.	F
6,7,8,9,10	The system should prompt the use of technologies to improve the efficiency of making the building firesafe in terms of both initial costs and whole life running costs.	NFP

In this case, the key for stakeholder identification numbers is as follows:

1 = enforcement authority, 2 = project team, 3 = fire engineer, 4 = architect, 5 = insurer, 6 = building owner, 7 = regulatory authority, 8 = fire industry, 9 = construction industry, 10 = government.

Table 23: HRM stakeholder specification

Operational Requirement	To provide a globally accessible method of formulating and evaluating fire strategies considering national requirements together with a broader consideration of objectives requirements and threat analysis
Functional Requirement	1. The system should make use of a consistent framework. Ideally, all submitted fire strategies will follow the same format, making both preparation and auditing easier. 2. The concept should ensure that all owner/occupier needs are appropriately met. The system should ensure that all stakeholder objectives and concerns can be fed into the methodology. This should consider the eventual use of the building.
Non-Functional Performance (NFP) Requirements	1. The auditing process for holistic fire strategies should be made as straightforward and unambiguous as possible. 2. The system should promote the best fire engineering knowledge. This could include a form of online code requirement search function based upon building profile and geographical region. 3. The system should require less time in assessing fire strategies than is currently required. 4. The system should prompt the use of technologies to improve the efficiency of making the building firesafe in terms of both initial costs and whole life running costs.
Non-Functional System (NFS) Requirements	1. The system should provide an audit trail for decisions made: Automatic logging for decisions made by all parties.
Non-Functional Implementation (NFI) Requirements	1. The process should recognise national regulations and codes. There should be built in method to allow fire strategies to follow national requirements and then amend as appropriate.

From Table 22, a separate table can be formulated by redefining the requirements into the Holistic Requirements Model for operational, functional, and non-functional requirements. This is shown in Table 23. The Table describes the specification for the holistic fire strategy proposal. The Stakeholder assessment model covered in this Chapter provides an interesting method of building up a specification for a system based upon a concept, which, in this case, is Holistic Fire Strategies.

The next chapter provides an overview of how the concept can be set up and operated.

14 An introduction to *igni.online*

An idea will stay just an idea unless it can be made available to the community.

Ype of building

Holistic Fire Strategies was always envisaged as an online methodology. This will allow it to be effortlessly accessible to all parties, wherever they are. It will also allow for continuous improvement. The website should follow the aims from stakeholder requirement analysis as identified in the previous chapter.

The concept will allow fire engineers working on a new project to access the website. A domain has been reserved for this application. The domain *"igni.online"* was taken from the Latin "ignis" and slightly modified – the "s" is removed. The ".online" tag denotes that this is an online application. Note that ".com" or another tag is not required. When writing this book, the website had been initially designed, although the format and operational structure of the site are not included in this book. The following description will provide an outline.

We should begin with those who will initially use the application. This will be the project management team overseeing the design and construction or modifications to the building. They may typically include the services of fire engineers within their team. Alternatively, a retrospective fire strategy may be required for an existing building when only fire engineers will be required. The process for a project is shown in Figure 23.

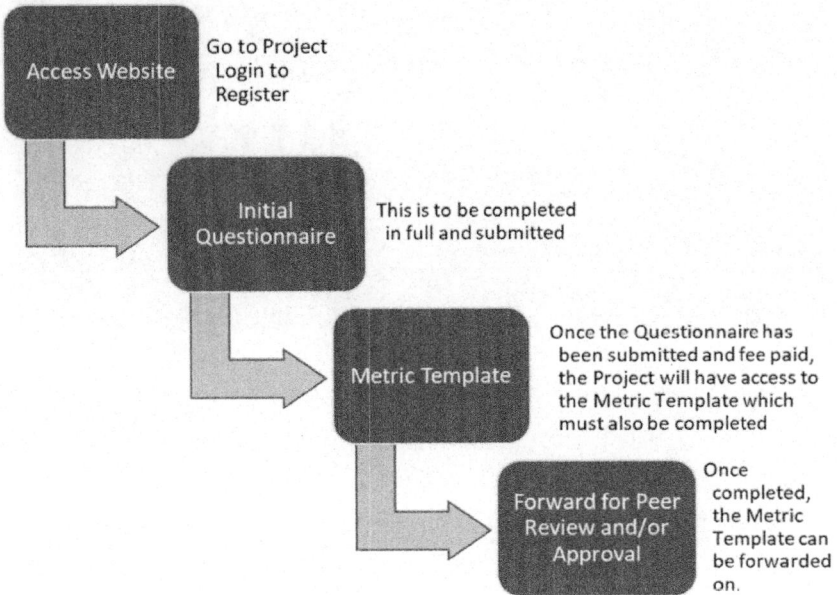

Figure 25: igni.online process

Accessing the website

Every new project will go to the registration page from which they will be given a unique registration login name. They will also be able to choose a password. They will be asked to complete an initial questionnaire.

The initial questionnaire

The initial questionnaire will be designed to filter out essential information, the most important of which is the location of the building as well as the use of the building – its profile. This can help focus the system on the types of code that will naturally be applied. Initially, the system may require the inputter to enter the relevant fire codes. However, it is easily conceivable that the system will automatically provide the user with the relevant codes based on the region and building type. An example of the questionnaire is shown in Table 24.

Table 24: The initial questionnaire (sample)

Name and location of the Project:	The name of the building being constructed and its address.
The primary purpose of the building:	This may include a list of tick boxes to allow consistency of data entry for different building profiles.
A description of the building profile:	General aspect of the building – its height, number of storeys including basements, construction materials, floor areas.
A description of the occupancy profile:	This will cover the numbers and types of occupants intended to occupy the building. For example, in the case of a hospital, it will include high dependency occupants.
A description of the internal processes:	The known and perceived uses of the building, including special risks, will be detailed here.
The relevant national codes to be applied:	The fire engineer will identify all the codes that will be applied to the fire engineering of the building, including prescriptive or performance-based methods.
The relevant enforcement authority and region:	A drop-down list of those enforcement authorities who will accept Holistic Fire Strategies will be included. It will be given in two parts –the country in which the building is located will filter those agencies within that country that support the scheme.

Once the form has been completed, there will be a fee charged. When this has been paid, then the questionnaire can be submitted. Finally, the application will be reviewed and, if acceptable, the project will be opened. At this point, the project will have access to the Metric Template.

The Metric Template

The metric template is the bones of the application. It is where all the information is entered, interrogated, and stored. A fully completed metric template will provide all the necessary information to provide a fully detailed and relevant fire strategy for the building in question. Figure 24 illustrates how the metric template may appear on the website. Note that the idea will be that the applicant will complete each of a series of "boxes", each covering a different aspect of the fire strategy.

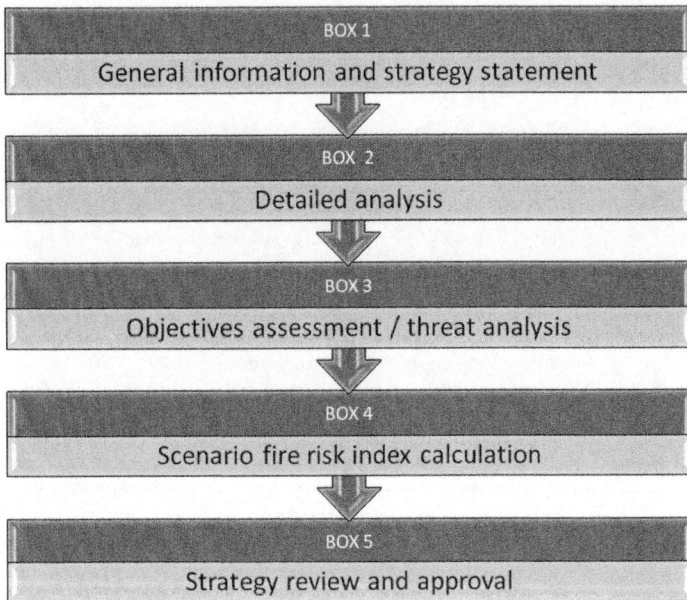

BOX 1
General information and strategy statement

BOX 2
Detailed analysis

BOX 3
Objectives assessment / threat analysis

BOX 4
Scenario fire risk index calculation

BOX 5
Strategy review and approval

Figure 26: The metric template

Those entering the information will click on a box and be taken to a series of relevant questions. Once the box has been completed, then it will be possible to move onto the next box, and so on.

One element that has been considered is the ordering of the boxes to provide, as far as possible, a seamless and complementary methodology. It is just possible that, as the template is developed, the box order may change. At this stage, the following order is thought to be most relevant.

Box 1: General Information and strategy statement

The person(s) entering the information will state the objectives for the fire strategy. The statement will include the following:

a) The title of the project with a brief description.

b) The stakeholders involved in the strategy.

c) The key objectives for the fire strategy. The input will be tested by box 3.

d) The fire safety legislation and codes applicable. Note that the box may prompt the relevant codes automatically based upon the building profile and its geographical location.

e) The approach taken – whether a prescriptive or performance-based approach is to be used. Limitations on the approach should also be stated (e.g., the fire detection systems will be designed to a prescriptive standard whilst the means of escape generally is based upon performance objectives).

f) Additional information such as assumptions made and performance requirements.

g) Programme for the design and construction of the building (where available).

Box 2: Detailed Analysis

This Box is a significant part of the metric template. The box will take the applicant through every single aspect of the fire strategy via a set of questions covering every element of a fire strategy in sufficient detail. The questionnaire for Box 2 will be split up into two parts. The following provides some guidance.

PART 1: Building design and use. This will include the following:

Building Characteristics: The characteristics of the building in terms of its location, height, construction materials, etc. Special features such as atriums, voids, basements, etc., should be stated.

Building plant and engineering services: The proposed building plant and engineering services should be stated, including the identification of areas susceptible to a fire in terms of location and separation from other risk areas. Furthermore, could access to critical services in the event of fire be compromised?

Building processes: All buildings contain a set of processes. These may be tangible and obvious or more subtle and part of the environment of the building. Key processes may include IT systems, manufacturing lines, storage facilities and equipment, chemical processes, etc. Similarly, the handling of people at airports and railway stations could be regarded as a building process.

Occupancy characteristics: The characteristics of each of the occupancy groups within any building will impact the overall life safety fire risk. Special groups such as sleeping risks, mobility-impaired persons, etc., should be stated here.

Proposed fire safety management strategy: Details of the level of fire safety management systems will be provided here.

PART 2: Fire safety and engineering design criteria

External Spread of Fire: The building's external construction, location, and potential exposure from fires from neighbouring buildings is a crucial risk factor. The exposure is based upon these factors:

a) Use of materials and attachments in the make-up of the external fascia of the building, including cladding types, insulation, protection of any cavities and the use of balconies and decorative features.

b) Distance from the neighbouring buildings and the level of potential exposure within the building from a neighbouring fire. Considerations will include the percentage of the unprotected area of the outer profile of the building (such as glazing).

c) Any risks of exposure via roofs.

Internal Spread of Fire and Smoke: Means should be included to control the spread of fire and smoke around the building. This is to ensure that persons can escape from the building without being exposed to a fire. Furthermore, restricting smoke and fire from spreading provides a better opportunity for firefighters to extinguish a fire at the source.

Control of fire and smoke around the building can be achieved by passive means (such as fire compartmentation) and active means (fire and smoke dampers, control of smoke ventilation and extract systems, etc.) Furthermore, it should be ensured that the internal linings of the building are chosen to minimise the spread of fire.

Fire compartment lines, fire doors, shutters, etc., including their fire-resistance rating, should be provided on a relevant set of drawings for the building.

The metric template will allow the uploading of drawing files to support the fire strategy. The questionnaire will cover the following:

a) Fire-resistance rating of structural elements.

b) Fire-resistance rating of floors and ceilings.

c) Internal fire-resistance rating (in terms of at least integrity and heat insulation) of:

   ✓ Primary / secondary compartments (where used to divide parts of the building internally).

   ✓ Escape staircases and lobbies (and special lifts for escape where required).

   ✓ Escape corridors.

   ✓ High hazard rooms.

   ✓ Special-use rooms (e.g., hospital operating theatres).

   ✓ Refuges.

   ✓ Other stated areas such as the internal parts of residential apartments.

d) Use of fire doors, shutters, glazed elements and fire and smoke dampers.

e) Smoke control provisions:

   ✓ Stated purpose(s) and performance objectives.

   ✓ Use of smoke curtains

   ✓ Use of passive or manually controlled ventilation systems in escape staircases, lobbies, internal corridors, etc.

   ✓ Use of manual and automatic smoke extract systems. The method(s) of activation will need to be stated, such as operation via fire detection systems.

✓ Use of pressurisation systems for escape staircases, special rooms, etc.

✓ Use of other forms of smoke control.

f) Automatic fire suppression systems:

✓ Stated purpose(s) and performance objectives.

✓ Use of main space (or total flooding) protection systems (such as sprinkler systems) and the extent of their application (e.g., throughout the building or for specific areas)

✓ Use of local application systems such as for the protection of specific rooms or pieces of equipment. The areas or equipment protected should be specified.

✓ The type of suppression agent should be specified, such as water fog systems, foam systems, chemical extinguishing systems, inerting extinguishing systems, etc.,) as well as the means of activating the systems.

g) Manual first aid fire extinguishing systems:

✓ Use of portable fire extinguishers.

✓ Use of hose reels.

✓ Use of specialist first aid fire extinguishing systems.

h) "Trade-off" provisions. Note that, in more performance-based approaches, one element of fire and smoke control can be traded off against another. For example, the use of sprinkler systems to allow for a reduction in fire

resistance ratings for fire compartments.
Smoke extract systems could allow for
increased escape time or travel distance by
maintaining the smoke layer above head height
for a design period. The conditions should be
stated here.

Evacuation and Means of Escape: There are two key
objectives to ensure successful evacuation:

- A means to discover a fire and raise the alarm
  – in good time.

- Appropriate means provided for all persons to
  safely evacuate the building, or they remain in
  or move to, a place of relative safety from where
  they can be assisted with their escape, if
  necessary.

The key criteria for this are as follows:

a) The method for detection of a fire, whether
   manual and/or automatic means. The type(s) and
   extent of fire detection systems should be stated.
   Any special detection system requirements
   should also be included, such as monitoring
   specific risks such as cable tunnels, electrical
   plant, oil and gas containers, special processes,
   etc.

b) The method(s) used for audible warning of a fire.
   The information relevant to this includes the
   types of warning methods (such as sirens, voice
   alarm, public address, etc.), how the warnings
   will be initiated, the extent of their use, and any
   special provisions, such as coded warnings and
   any required delays to allow for investigation
   times before initiating evacuation.

c) Additional methods for warning of a fire. This may
   include visual alarm systems, personal alert

systems using vibrating devices or mobile telephones.

d) The need to automatically operate or activate other systems to aid evacuation, such as hold-open devices for fire doors, fire dampers to maintain the integrity of passive structures, the operation of smoke control systems, etc.

e) The type of evacuation chosen:

> ✓ Total or simultaneous evacuation: This is where all persons are simultaneously evacuated from the building, possibly automatically initiated following the detection of a fire condition. The primary consideration here is if the means of escape can cater for the maximum number of persons to evacuate safely and within the specified evacuation period.

> ✓ Phased evacuation: This is where persons are evacuated in stages. An evacuation alarm may be given in the designated areas or floors whilst an alert alarm is given in other areas. In this case, the passive and active fire protection arrangement should ensure that those remaining in the building are separated from the fire for the entire duration of the waiting period and up until they reach a place of designated safety.

> ✓ Managed/coded evacuation: This is where trained personnel assist with the evacuation process and may involve using a coded alarm message to alert relevant building personnel. This strategy will give time to investigate the incident before initiating evacuation.

- ✓ Manual evacuation: This is where evacuation is controlled by building personnel only and may vary depending on the incident. Messages may be given via a PA system. This method relies on suitable training of key personnel and detailed approved procedures and parameters.

- ✓ "Stay put" evacuation strategy whereby persons are requested to stay in place during a fire until they are advised to evacuate. This strategy may be found in residential apartment blocks.

f) For horizontal means of escape, the criteria for the following should be stated:

- ✓ The maximum horizontal dimensions for a fire zone. This may require consideration of the maximum number of people expected within that zone.

- ✓ If progressive horizontal evacuation is used (whereby vertical evacuation is not practicable, such as is the case with hospitals), then the special provisions should be stated.

- ✓ The maximum sizing of one-way and multi-way travel distances to a place of safety. Note here that relaxations due to the risk profile or the use of systems should be explicitly stated.

- ✓ The sizing of escape corridors and exit doorways.

- ✓ Any special and additional horizontal arrangements should be specified.

g) For vertical means of escape, the criteria for the following should be stated:

   ✓ Access criteria to reach the escape staircases.

   ✓ Numbers and dimensions of escape staircases, including their final exit points from the building. The facilities should be closely related to the type of evacuation chosen.

   ✓ Use of lobbies to separate escape staircases from risk areas.

   ✓ Additional facilities to maintain integrity or tenability of escape staircases such as smoke ventilation, extract, and pressurisation systems.

   ✓ Use of lifts for evacuation purposes.

h) The provision of refuges for where persons are not able to directly evacuate from the building. Considerations include:

   ✓ Sizing and location of refuges.

   ✓ Use of fire telephone systems to allow voice communication.

   ✓ Means to assist evacuation from refuges, including equipment to assist in the evacuation of mobility impaired persons.

i) Alternative specification criteria used to determine means of escape, such as the use of performance-based fire engineering. This may include fire and evacuation modelling to evaluate the available safe escape time (ASET) and the required safe escape time (RSET). The idea here is that the available time is sufficiently greater than the time required for persons to escape.

j) Means to illuminate and point out the direction of the escape paths by emergency lighting and active or passive fire signage.

k) Special arrangements such as escape provisions for atriums should be stated.

Firefighting arrangements: A range of logistical and support issues will be relevant, and these may include:

a) Means of contacting the fire and rescue services.

b) Criteria for attendance time to the site, particularly where the building is in a rural area.

c) Access arrangements to the building, including suitable vehicular access arrangements.

d) Access for firefighters into the building, to upper or lower levels, and other risk areas.

e) Availability of fire hydrants near and around the building.

f) The provision for firefighting mains water supplies (wet or dry risers) in and around the building and on all levels.

g) The provision of firefighter lifts.

h) The provision for smoke venting, extract, and clearance facilities to allow for firefighting at all levels, particularly in basement areas.

When the questionnaire has been completed, the basis of the fire strategy will be available. It can then be printed out as a draft document – possibly in a preferred national format. However, the holistic nature of the fire strategy has still yet to be developed.

Box 3: Objectives Setting / Threat Analysis

The purpose of this box is to allow the undertaking of an objectives setting and threat assessment. This methodology is covered in detail in earlier chapters.

Although the proposed evaluation would seem complex and laborious, the intention is to formulate algorithms adapting the analysis contained in this book. The algorithms will then provide the chosen scenarios with risk ratings leading to the additional control measures required. These additional measures will then automatically update the fire strategy with additional or alternative provisions.

At this point, the fire strategy has been completed. Evaluation of additional objectives and threats may have led to additional inclusions within the fire strategy.

Box 4: Scenario fire risk calculation

At this point, the fire strategy can be tested against an idealised baseline fire strategy as described in a previous chapter. The Fire Strategy Risk Index will provide those involved in the project sufficient comfort that the strategy is fit for purpose. That is, if the residual risk is found to be less than the baseline fire strategy. If the risk is higher, then the output of this Box will provide suggested amendments to the strategy. The user will then be guided back to Box 2.

Box 5: Strategy review and approval

Box 5 will allow the review of the completed fire strategy in a format relevant to the country, region, and naturally, the appropriate enforcement authority. For instance, if the strategy is formulated for a UK based building project, it would be possible to review the strategy in a format akin to the UK Building Regulations – Approved Document B. At this point, relevant drawings, the results of fire and evacuation modelling, references, and so on can be included, and any other supporting information provided as appendices.

Once all the earlier boxes have been completed, the next stage will be to pass for approval. The system users will have two options. The first option is to send the completed template for Peer Review. The second is to forward directly to the relevant enforcement authority.

The peer-review process will involve an independent assessment by an appropriately experienced and qualified fire engineer. They will revisit each element of the metric template and judge if the template has been completed correctly. They will also assess whether the correct assumptions have been made. Once the review has been completed, a separate report will be sent back to the project. This report will highlight the corrective actions that are deemed necessary. The peer-review process will be chargeable to the project.

The completed template can then be forwarded to the relevant enforcement authority. The submission must be accurate and complete. Each enforcement authority will eventually have a measurement process designed explicitly for them. The process should provide them with an unambiguous, concise and conclusive fire strategy. The whole methodology should require a small percentage of their resources compared to the current paper-based fire strategies. It is also likely that they may insist that the submission goes through the peer-review process. This will give them additional confidence that the fire strategy is fit for purpose.

The strategy can be viewed online or printed in a known and consistent framework to allow easy verification. Should the enforcement authority approve the strategy, they will submit an "approved" notice directly via the website. Should they query the strategy, there will similarly send a "rejected" notice which will come with comments.

By the time this book is published, *igni.online* should be up and running as a sample site but is unlikely to be operational in the near future.

15 Conclusions and future development of the concept

An idea, if it is a good idea, should be pricked and poked to see if it stands up to scrutiny.

This book was completed during the worldwide Covid19 pandemic of 2020 and 2021. As I was writing the book - in semi-lockdown in Sitges, Spain, I had a realisation. The world was about to change even if we aim to return to a kind of new normality. Every aspect of life we took for granted at the end of 2019 will become a cherished memory for many. For some, though, it became a positive change for good. For many months, the skies had cleared, wildlife had been able to roam free again, and the polluting industries had stopped for a time – time for the earth to heal just a bit. Terms like "the new normal" were, and are, bandied around.

Could the "new normal" also apply to the fire industry? Is what we had taken for granted as a standard methodology, for so long, still justifiable?

I had noticed that there were certain parallels with the proposals detailed in this book:

Objectives. It was evident that human objectives were quickly re-evaluated when we realised we could lose precious members of our family or friends. A major crisis has suddenly prompted us to reassess our actual goals. There are more important things to life than a big house or a nice car.

Threats. Many countries were unprepared for the virus despite warnings from the medical community. The results were devastating. Our whole treatment and understanding of threats have revealed weaknesses. A more robust method of threat analysis is required.

A global approach? The virus did not recognise national boundaries. The Covid19 strategy adopted by countries revealed stark variations in the infection rate, death rate, and the length of lockdown. It was evident to many that the only way a pandemic can be effectively handled is by adopting a global approach rather than a national one.

This book follows on from a "eureka" moment in January 2017. It has taken time to get to this point. Nevertheless, the work is a long way from anywhere near over. Even though many of my ideas now finally appear in print, please accept them as a first pass. New concepts require a degree of honing. This book may need to be reviewed and revised at some stage. My search has only just got started!

Even though the concept behind holistic fire strategies is yet to be up and running, there are already ideas for future improvement.

Help and assistance for those <u>preparing</u> the fire strategy

What if fire engineers were given help and advice as they progress through the boxes? For instance, for every question clicked, a help tab automatically appears. This help tab will provide relevant information based on the code applicable to that building type and location. This would require the input of a vast range of data covering every national code with a sub-set for every type of building. An enhanced metric template will guide the applicant to complete the strategy with the vital information at hand. This offering will further improve consistency as well as increase the value of the application.

Help and assistance for those <u>approving</u> the fire strategy

What if the concept redlined those areas of a fire strategy that may differ from the principal codes, making auditing even easier and more focused? The original idea is that enforcing authorities can readily access each project to get a real-time understanding of project progress. Similarly, the system could proactively prompt the enforcement authorities as progress is made in a specific project under their jurisdiction.

Continuous learning from major fires

The early chapters of this book point out that national life safety legislation was often formulated after a major fire. With modern media, the news of a new fire tragedy is revealed to all of us in minutes. In some cases, the causes and consequences are also known in a matter of hours and days. What if every event is flagged into the igni.online system as lessons learned? Quite possibly, the information can automatically update, say, a high-rise office building project if there are specific lessons from a recent actual case of a high-rise office building fire somewhere in the world. A separate library of these events could be maintained and accessed via the system as an online library.

Interconnecting with BIM

Building Information Modelling (BIM) was described in an earlier chapter of this book. This involves the use of intelligent 3D models in the design of buildings. It enables a complete audit trail to be established for every element of the planning, design, build, operation, and maintenance phases. What if the holistic fire strategy concept could be directly made an integral part of the BIM process? Architects and building teams would have the additional assurance that the design is highly likely to be accepted from at least the perspective of fire safety.

The next steps

I would like to thank everybody who has read this book – especially to this point. I will be refining the ideas as time goes by, but this book is very much my *flag in the ground*.

The search continues!

ABOUT THE AUTHOR

Paul is best known for his work in developing the concept of robust fire strategies around the world. He wrote British Standard Specification PAS 911 in 2007 and, in 2013, authored his much-acclaimed book "Fire strategies - strategic thinking". Paul formed his business Kingfell in 1995 and grew it to a multi-million fire one-stop-shop before changing direction. He is now a Founding Partner of Fire Cubed LLP (www.firecubed.com) and co-founder of an Authorising Engineer business serving the NHS - AdashE (www.authorising-engineer.com)

In his earlier career, Paul was Head of Fire Engineering for London Underground. Before that, he worked for the Loss Prevention Council and Fire Offices' Committee in the UK. He is a Chartered Fire Engineer, Member of the Institute of Fire Engineers, and liveried member of the Worshipful Company of Firefighters. Paul was named in the top 5 of most influential persons in fire safety by IFSEC Global for 2018. He received his PhD in Civil Engineering in June 2021.

References

[1] Porter, Michael E: *"Competitive Strategy: Techniques for Analyzing Industries and Competitors"*. Boston Free Press, 1998.

[2] Bryant, Paul: *"Fire Protection Strategies - Parts 1 and 2"* Published in Fire Surveyor Journal (Hertfordshire 1996)

[3] British Standard Specification PAS 911: 2007: Fire strategies – guidance and framework for their formulation

[4] British Standard BS 9999: 2017: Fire safety in the design, management and use of buildings.

[5] British Standard BS 7974: 2019: Application of fire engineering principles to the design of buildings. Code of practice.

[6] UK Home Office fire statistics – UK Government 2017

[7] NFPA (US) – *Fire loss in the United States during 2017* NFPA 2018

[8] NFPA (US) – *Fire loss in the United States during 2017* NFPA 2018

[9] NFU Insurance. nfu-say-80-of-businesses-fail-within-18-months-of-a-major-fire-incident. https://www.fire-magazine.com. [Online] June 2019.

[10] Safety Management UK. [Online] [Cited: June 2019.] http://safety-managementuk.com/70-percent-fail.php.

[11] Bryant, P: *"Fire Strategies - Strategic Thinking"* 2013. Amazon Createspace

[12] Confidential discussions with members of regional fire and rescue services.

[13] British Standard Specification PAS 911: 2007: Fire strategies – guidance and framework for their formulation.

[14] Bannister, Adam. *The 5 Fire Safety Failures that Fuelled the Great Fire of London.* August 2016: IFSEC Global News site.

[15] Fire Magazine (online) https://www.fire-magazine.com/woolworths-inferno-forces-legislation-change July 2010

[16] UK Building Regulations 2010. Approved Document B: Fire Safety: 2013. UK Government

[17] British Standard BS 9999: 2017: - Code of practice for fire safety in the design, management and use of buildings

[18] British Standard BS 9997:2019: Fire risk management systems. Requirements with guidance for use.

[19] British Standard Draft for Development DD240: Fire safety engineering in buildings. Guide to the application of fire safety engineering principles. Withdrawn, 1997.

[20] British Standard BS 7974: 2019: Application of fire safety engineering principles to the design of buildings. Code of practice.

[21] Hosch W. L. Navier-stokes equation. Brittanica. [Online] Encyclopædia Britannica., 2020. [Cited: June 2020.] https://www.britannica.com/science/Navier-Stokes-equation.

[22] Rodrigues, Eduardo EC, Rodrigues, Joao PC and Filho, Luiz *Comparative study of building fire safety regulations in different Brazilian States*. CP. 10, Rio Grande do Sol, Brazil: Elsevier, 2017, Vol. Journal of Building Engineering. 102-108.

[23] International Standards Organisation (ISO) *https://www.iso.org/*

24 ISO 6182:2020: Fire protection. Automatic sprinkler systems.
25 https://ifss-coalition.org/
26 Chinese Government Websites:
http://www.mohurd.gov.cn/wjfb/201508/t20150829_224333.html
http://www.mohurd.gov.cn/wjfb/201805/t20180509_235971.html
27 Hong King Code: https://www.bd.gov.hk/doc/en/resources/codes-and-references/code-and-design-manuals/fs2011/fs2011_full.pdf
28 NFPA 101, Life Safety Code: 2021
29 NFPA 101A: Guide on Alternative Approaches to Life Safety: 2022
31 Building Authority of Hong Kong: Code of Practice for provision of means of escape in case of fire 1996 https://www.bd.gov.hk/doc/en/resources/codes-and-references/code-and-design-manuals/MOE1996_e.pdf
32 Jianga, Jian *et al* : *Fire safety assessment of super tall buildings: A case study on Shanghai Tower* Elsevier Case Studies in Fire Safety Volume 4, October 2015, Pages 28-38
33 EN 13501-1:2018: Fire classification of construction products and building elements. Classification using data from reaction to fire tests. (Also available as a BS).
34 IS 2189: Selection, installation, and maintenance of automatic fire detection and alarm systems – Code of Practice -Fourth Revision
35 NZS 4512:2021 Fire detection and alarm systems in buildings
36 BS 5839-1:2017: Fire detection and fire alarm systems for buildings. Code of practice for design, installation, commissioning and maintenance of systems in non-domestic premises
37 NFPA 72, National Fire Alarm and Signaling Code: 2019
38 IS 3844: Code of Practice for Installation and Maintenance of Internal Fire Hydrants and Hose Reels on Premises (First Revision) : Bureau of Indian Standards (BIS): Publication date 1989
39 NFPA 14, Standard for the Installation of Standpipe and Hose Systems: 2019
40 NFPA 1, Fire Code: 2018
41 NFPA 1710, Standard for the Organization and Deployment of Fire Suppression Operations, Emergency Medical Operations, and Special Operations to the Public by Career Fire Departments: 2020.
42 Internet source provided by Wikipedia -
https://en.wikipedia.org/wiki/Fire_protection_engineering
43 ^ Rowland G. M. Baker, Fire Insurance Wall Plaques Walton & Weybridge Local History Society, Paper No 7, 1970
44 The Fire Offices' Committee was disbanded in 1985. Its services were transferred to the Loss Prevention Council (UK) and the Fire Protection Association (UK)
45 Factory Mutual Global www.fmglobal.com
46 Underwriter's Laboratories www.ul.com
47 Website https://www.lostcolleges.com/armour-institute
48 Institution of Fire Engineers Website https://www.ife.org.uk/About

[49] Society of Fire Protection Engineers Website
https://www.sfpe.org/page/about

[50] Institution of Civil Engineers website. What is civil engineering? [Online] 2020.
[Cited: August 2020 .] https://www.ice.org.uk/what-is-civil-engineering &
What is Civil Engineering. ICE. 15 May 2017.

[51] John, Jesse "Societal Engineering: What is it, and Why is it Valuable?" May 21,
2017 · https://medium.com/societalengineering/societal-engineering-what-is-
it-and-why-is-it-valuable-906be680d108

[52] Institute of industrial and systems engineers (US)
https://www.iise.org/details.aspx?id=282

[53] Environmental Engineering (UK) Ltd http://www.environmental-
engineering.com/en_GB/about-us/

[54] Academy of environmental engineers and scientists (USA)
https://www.aaees.org/careers/

[55] Risk Engineering website https://risk-engineering.org/

[56] Stanford University (California, US)
https://cheme.stanford.edu/admissions/undergraduate/what-chemical-
engineering

[57] Marriam Webster https://www.merriam-
webster.com/dictionary/mechanical%20engineering

[58] Wang YC. Tensile Membrane Action and the Fire Resistance of Steel Framed
Buildings. Melbourne, Australia : Fire Safety Science- 5th International
Symposium, 1997.

[59] Bailey C. Construction industry slashes cost of fire protection in steel frame
buildings. University of Manchester -Department of Mechanical, Aerospace and
Civil Engineering. [Online] 2020. [Cited: June 2020.]
https://www.mace.manchester.ac.uk/research/impact/ref-
impact/construction-industry-steel-frame-buildings

[60] The University of Manchester (Civil Engineering) website. Infrastructural fires
(Case studies/historic fires). [Online] 2019. [Cited: December 2019.]
http://www.mace.manchester.ac.uk/project/research/structures/strucfire/Cas
eStudy/HistoricFires/InfrastructuralFires/mont.htm.

[61] Troisi R., Alfano G. Towns as Safety Organizational Fields: An Institutional
Framework in Times of Emergency. Sustainability. December 2019, Vol. 11 (24),
7025.

[62] Dent S. Fire Protection Engineering and Sustainable Design. 2010, 46, 10–16.
Available online:. sfpe.org. [Online] 2010, 46,10-16.
https://www.sfpe.org/page/2010_Q2_1.

[63] Pilkington Glass. Pilkington . [Online] 2020. [Cited: 29 May 2020.]
https://www.pilkington.com/en-gb/uk/architects/glass-information/

[64] Digital School website. History of BIM. Digital School. [Online] 2020.
https://www.digitalschool.ca/history-bim.

[65] Strömgren M. BIM and Fire Safety: Raising the bar for quality. Firesafe Europe.
[Online] 2020. [Cited: July 2020.] https://firesafeeurope.eu/bim-and-fire-safety/

[66] International Standards Organization: ISO 30001: 2018: Risk management

[67] British Standard Specification PAS 79:2012: Fire Risk Assessment. Guidance and a recommended methodology

[68] British Standard PD 7974-7:2019: Application of fire safety engineering principles to the design of buildings. Probabilistic risk assessment

[69] SFPE Handbook of Fire Protection Engineering - Fire Risk Indexing. Watts Jr, M John: Society of Fire Protection Engineers, 2016, Vol. SFPE 82.

[70] American Institute of Chemical *Engineers Tools for Making Acute Risk Decisions with Chemical Process Safety Applications.* New York:, 1994.

[71] Rosenblum, G.R. and. Lapp, S.A *The Use of Risk Index Systems to Evaluate Risk in Risk Analysis: Setting National Priorities*, Proceedings of the Society for Risk Analysis, Society for Risk Analysis, 1987.

[72] SE UK Health and Safety Executive. *Alarp at a glance. Risk Theory. [Online]* (UK). http://www.hse.gov.uk/risk/theory/alarpglance.htm.

[73] Watts Jr, M John SFPE Handbook of Fire Protection Engineering - Fire Risk Indexing.. s.l. : Society of Fire Protection Engineers, 2016, Vol. SFPE 82.

[74] ISO 16733-1:2015: Fire safety engineering — Selection of design fire scenarios and design fires — Part 1: Selection of design fire scenarios

[75] Hadjsophpcleous and Mehaffey, SFPE Handbook of Fire Protection Engineering - Fire Risk Indexing.. s.l. : Society of Fire Protection Engineers, 2016, Vol. SFPE 82.

[76] Hadjisophocleous G.V., Mehaffey J.R. (2016) Fire Scenarios. In: Hurley M.J. et al. (eds) SFPE Handbook of Fire Protection Engineering. Springer, New York, NY https://link.springer.com/chapter/10.1007%2F978-1-4939-2565-0_38#citeas

[77] Gaskin J and Yung D, "*Canadian and U.S.A. Fire Statistics for Use in the Risk-Cost Assessment Model*," IRC Internal Report No. 637, National Research Council of Canada, Ottawa, (Jan. 1993).

[78] New Zealand Building Code Clauses C1-C6 Protection from Fire 2013.

[79] Xin, Jing, Huang, Chongfu Fire risk analysis of residential buildings based on scenario clusters and its application to risk management, Fire safety journal 62 (2013) 72-78, Elsevier

[80] NFPA 551: 2019: Guide for the evaluation of fire risk assessments.

[81] Kingfell Guide KF912: Crisis management planning (Kingfell Ltd 2013)

[82] Taken from various internet news reports

[83] Guardian Newspaper website: https://www.theguardian.com/uk-news/2018/mar/16/towers-with-grenfell-style-cladding-at-risk-of-arson-and-terrorism

[84] UK Home Office Report "Detailed analysis of fires attended by fire and rescue services, England, April 2017 to March 2018. Statistical Bulletin 17/18-6th Sept 2018

[85] The National Journal. [Online] https://www.thenational.ae/uae/cigarette-caused-dubai-s-torch-tower-fire-1.624072.

[86] WHIO TV. news/traffic-accident-leads-massive-warehouse-fire. whio.com. [Online] https://www.whio.com/news/traffic-accident-leads-massive-warehouse-fire/yum9lP5wm7DE52J8GcOcaJ/.

[87] McSheffrey, B. *Fire Extinguishers Extinguish an Estimated 5.32 Million Fires in US in 2010* -En-Gauge Fire and Life Safety Blog 17th Oct2 011
http://www.engaugeinc.net/life-and-fire-safety-blog/fire-extinguishers-extinguish-an-estimated-532-million-fires-in-us-in-2010
[88] Brzezińska D, Bryant P, 2018, *Buildings Fire Protection Strategies*, Lodz University of Technology, Lodz, 209 p.
[89] Brzezińska D, Bryant P, 2018a, New Anglo-Polish Methodology for Fire Strategies Evaluation, 15th International Conference on Fire Science and Engineering, Interflam 2019, Royal Holloway College -University of London, 1st - 3rd July 2019
[90] Gretener M, 1973, Evaluation of Fire Hazard and Determining Protective Measures, Zurich: Association of Cantonal institutions for Fire Insurance (VKF) and Fire Prevention Service for Industry and Trade (BVD).
[91] Brzezińska D, Bryant P, 2018, Buildings Fire Protection Strategies, Lodz University of Technology, Lodz, 209 p
[92] Burge, Stuart, "The Systems Engineering Tool Box" 2006 – website
https://www.burgehugheswalsh.co.uk/Uploaded/1/Documents/Holistic-Requirements-Model-Tool-v1.pdf

Printed in Great Britain
by Amazon

64969286R00139